Aaron Stevens Hayward

VITAL MAGNETIC CURE

Aaron Stevens Hayward

VITAL MAGNETIC CURE

ISBN/EAN: 9783741121920

Manufactured in Europe, USA, Canada, Australia, Japa

Cover: Foto ©Angelika Wolter / pixelio.de

Manufactured and distributed by brebook publishing software (www.brebook.com)

Aaron Stevens Hayward

VITAL MAGNETIC CURE

VITAL MAGNETIC CURE:

AN EXPOSITION OF

VITAL MAGNETISM,

AND ITS

APPLICATION TO THE TREATMENT OF MENTAL AND PHYSICAL DISEASE.

BY A MAGNETIC PHYSICIAN.

BOSTON:
WILLIAM WHITE AND COMPANY.
158 WASHINGTON STREET.
NEW YORK AGENTS:—THE AMERICAN NEWS COMPANY,
119 NASSAU STREET.
1871.

CONTENTS.

VITAL MAGNETIC CURE.

INTRODUCTORY.

FROM the remotest period of human history the power of mind over mind, and mind over matter has been known. In all ages it has to some extent, and in some form been illustrated; and there have descended to us in legitimate annals or traditional story, occasional glimpses or more complete accounts of its application in the treatment of disease, followed by the most marvellous results. With only the light of a materialistic philosophy to guide, the statements could not be regarded otherwise than marvellous, if not deemed absolutely incredible. Substantiated by competent and reliable testimony, yet contrary to general experience, and in opposition to known physical laws, they could only be explained by a belief in miracle. But since the boundaries of natural science have been vastly enlarged, it has come to be understood that such phenomena are neither unnatural, supernatural nor miraculous. Although at certain epochs their manifestation has been confined to a few persons, or known only in isolated instances, the extent to which they have been observed thus fluctuating from time to time, they have in latter years rapidly increased, so that in our day they are of no uncommon occurrence.

The Biblical history contains, both in the Old and New Testaments, numerous accounts of the cure of disease, or "healing" as it was termed, by the "laying on of hands." In the Apostolic times they were at first denied, for what could there be of value in the simple process of manipulation, capable of producing such results? Yet the statements were made with all the assurance of positive and undeniable fact. Positive testimony given by living witnesses, according to the record, proved the assertions to be true. Since the facts could no longer be denied, the motives of Jesus and his followers were impugned; and it was declared that the casting out of demons was accomplished only because Jesus was in league with the Prince of them. "He hath a devil." Thus also is it now, when a new fact or phenomenon occurs which is not readily understood, nor explainable in accordance with previous educational prejudices. An attempt is made to deny the facts. When the testimony is so strong and so easily accessible, that it would become mere stupid idiocy to reiterate their denial, the cry is "humbug," and no matter how good the deed, it is done through the instigation of the devil. But thinking men and women refuse to be satisfied with such a solution of a scientific problem. When a fact is self-evident, or established by reasonable testimony, they must receive it as a truth, upon its own merits; and truth is too precious to be cast aside, because they cannot always on the instant explain it. The study of a phenomenon will some time or other be rewarded by an understanding knowledge, and for this the honest and earnest student of nature can patiently wait.

The phenomena adverted to have been observed at different times and in different countries. Observers have experimented independently, and reached conclusions which in the main have been uniform. They have been brought before the world under different names, as Mesmerism, Animal Magnetism, Human Magnetism, Vital Magnetism, Psychological Power, Spiritual Influence &c. It has been reserved for later times to see the rapid spread of this curative power, and the general diffusion of knowledge concerning it.

It is a matter of common observation that some practitioners of the healing art have always been more successful than others, with no apparent advantage however at the outset, in acuteness of intellect, diligence of application to study, efficient culture, or extent of experience. It has generally been attributed to superior judgment, nothing having been credited to intuitional perception or psychometry: but the true reason of this difference is now beginning to be understood.

The treatises on Animal Magnetism have been complete, so far as relates to the control which one *will* can exercise over another; and the insensibility to pain during surgical operations, induced by the mesmeric process. The psychologic power has been fully and ably set forth in an admirable treatise entitled "Mental Cure" by W. F. Evans, recently published, which we cordially recommend to our readers.

The present treatise is based upon the author's observation and experience in the use of Vital Magnetism. It is intended to show the similarity, and the frequent identity of the gift of healing by the " laying on of hands " as described in the records of biblical history in former times, with events that are transpiring in the nineteenth century in our midst; and the harmony of the operation of the law which governed the ancient, with that which governs the modern phenomena. The process is a natural one, the present manifestations corroborating those of the olden times. It contains all the cases recorded in the New Testament, and many of those in the Old. Coming down nearer to our own time, a historical account is given of the experiments of Mesmer, Animal Magnetism as practised in the middle ages, the Report of the Commissioners appointed by the French Government, the opinions of many distinguished persons on the subject, a statement of occurrences taking place in the present generation, and the mode of applying Vital or Spirit Magnetism for the purpose of eradicating disease.

The principles upon which this practice is based are founded in truth, and have been so demonstrated to the satisfaction of critical investigators. The facts concerning its curative power have been realized by millions of persons in this and other countries within the last few years. It is resorted to by constantly increasing numbers as rapidly as prejudice is made to give way; and there can be no doubt that as soon as the mass of mankind are convinced of its utility, it will necessarily be brought into general use.

It has already been recognized by prominent medical men, who recommend it for its unquestionable efficiency. And among candid and philanthropic physicians, whenever opportunity is afforded for comparing its effects with the unavailing results of persistence in other modes of practice, doubtless the use of drugs will be gradually discarded, and the vital magnetic plan adopted. They will cultivate the latent power which they possess, aided by their acquired learning to a better understanding of the laws of life in Nature, and thus become more and more useful in their vocation. It will be only the lesser minds, of limited attainment's, whose bigotry and intolerance will bar the way to the investigation of truth.

It has been observed that "In all Science—throughout all Nature, the most invisible elements are the controlling forces. The physician deals with nice care the infinitessimal doses of the more subtle poisons, but the chemist cannot weigh the morbific air, which laden with fatal virus fells its thousands; and while material qualities are least controlling, and least to be feared, they are also least to be loved. Though we may handle and taste the crude forms of nature, we cannot fathom the fine essentials as exhibited through and in the deep mysterious fountains and reservoirs of the Infinite, created and prepared for the development and ultimate restoration and perfection of man. Thus in Medicine, there are substances, which to measure, weigh or understand their chemical nature, will not fully point out their therapeutic relation to diseased conditions. Such a substance or principle is Electricity or Magnetism. Its effects have been ascertained and its remedial powers partially known by experiments made when all other means have failed. Physicians of all schools have found that its place cannot be supplied in diseases of the Nervous Organism. Many contend that Nerve Force and Electro-Magnetism are identical, and that in depressed vitality or paralyzed nerve force the deficiency is actually made up; others, that its results are the effect of stimulation of nerve power. Be that as it may, its effect in relieving pain, removing paralysis and allaying disturbances of the nervous system, has become a settled fact in science, and this subtle agent is now one of the sheet anchors of the *Materia Medica*.

"The grand curative effect of electricity has been the result of application by ordinary Batteries, and in many cases by those uneducated in its use, and as a last resort when other means had failed, and vitality was expended, yet here this potent agent has left its signal mark and astonished by the material dose given. If such results can be produced under such unfavorable circumstances, when so little nerve force is left to act upon, and that too with imperfect forms of Batteries and in unskilled hands, what will be the force and curative power of this principle, when the current can be applied always with the same force to the entire surface and brought into contact with every function, organ and nerve; when the skilled hand of Almighty Power sends it forth from the caverns of his own labaratory "with no variableness or shadow of turning."

We are further assured from a source of high intelligence that "The application of magnetism, properly, not improperly, is destined to do away with all narcotics as remedial agents. They are at present an absolute necessity to human ignorance. But when ignorance shall have given place to knowledge, the veriest child will know how to use the magnetic powers with which every human body is endowed. You all hold within your grasp all the remedial agents that you have need of, but you do not know how to use them. Disease being an imponderable, it can best be treated by the application of an imponderable. That all-powerful force which you call magnetism, holds within itself the power to harmonize all the forces of the human body—to prevent disease. When disease, or inharmony, which is the same, has found an abiding place in the physical form, magnetism has the power to eradicate it, to overcome it; not only to subdue it, but to entirely overcome it; but the time for these things is not yet. You are standing, to-day, upon the threshold of this new science. It has always been with you, but because of its simplicity men and women have considered it of no acconnt. But the time is fast speeding when you will understand what disease is, and what is its remedy. You will also know how to apply the remedy. But you grow slowly, and you can grow no faster than the earth grows. Were we to be endowed this hour with infinite wisdom, it would be of small account to us, because we are not ready for it. We must grow up to a condition fit to receive it, to use it well, ere it can come to us. So humanity must suffer a while longer ere the angel of healing can come perfectly to your conscious lives and teach you what you so earnestly desire to know in the present."

By permission of the proprietors, we have in some instances availed ourselves of valuable information upon psychological questions, contained in the Message Department of the "Banner of Light," communicated through the instrumentality of Mrs. J. H. Conant.

In alluding to the diversity of magnetic endowment, we have refrained from mentioning the names of particular Magnetizers and Clairvoyants, in order that we might deal impartially with all. The treatise is intended to benefit all, by making known what can be accomplished by this remarkable power. It is a book for the skeptic and the honest inquirer, as well as these classes; and in fact for all who dare to think for themselves on these great living, vital truths. More might be said on the subject, for it is inexhaustible. Others should strive to improve upon our effort, giving to the world their highest and best thoughts, and the teachings of experience, leading to a correct understanding of life, both in the material and spiritual realms of existence.

THE GIFT OF HEALING.

Having been endowed with that peculiar power over the elements of diseased action, known in the Apostolic times as the "gift of healing," by which the sick are restored to health without the use of medicine; and having had considerable experience in the treatment of different forms of disease by its exercise, we have felt it a duty to communicate to the public the results of that experience. We have not been limited to personal experience, however, but in order to learn the value and extent of this most important method of treatment, have visited "healers" in all parts of the country, comparing the different phases of Vital Magnetism, and the various methods of its employment. It is important that those who assume to exercise the functions of magnetic physicians should be well informed in every thing that pertains to their office, that they may know what can be done by the vital magnetic force, and what is the limit of its application to the removal of disease, and the restoration of the sick to health. The more accurate is this knowledge, and the more extensively it is diffused, the more general will be the disposition on the part of invalids to seek its employment, and the better will it be, not only for its practitioners, but for all physicians and the public at large.

A variety of mechanical operations have been resorted to, for the purpose of arousing the different organs of the body to action, such as chafing, rubbing and pinching; and progressive physicians have advised such means, in the belief that their patients would be benefited thereby. Many of them have discovered that whatever of benefit has been thus derived, has depended more upon the magnetic power, electric force, or spirit power possessed by the operator, than upon friction, or other merely mechanical action.

Doubtless much can be done by the process of rubbing the patient, even if those who practice it are not liberally supplied with magnetic or electric force. Patients may benefit themselves also by that means, as well as by daily kneading manipulations, in preventing or warding off the approach of disease, and abating its force when an attack has already commenced.

In magnetic treatment the necessity of exposure in those physical examinations which are considered unavoidable in ordinary practice is superceded. The magnetizer soon discovers, whether intuitively or otherwise, the condition of the sufferer. He simply touches the patient and perceives, as it were by a glance of the eye, the condition of the blood's circulation, whether the vital forces are depressed, or there is an unhealthy fulness of habit, and deduces at once the need of electricity or magnetism, to bring about a state of equilibrium. The adaptation of the required force or power is vastly more essential in removing disease than the utterance of words, whether solemn or cheerful, which some are in the habit of using. We have found, furthermore, in the course of our investigations, that the adaptation of forces consists more in quality than in quantity. Thus delicate females in some instances can perform cures that seem wonderful, upon patients who, in their ordinary health, are much superior in muscular strength, and whose temperament is positive: and in the same way children have been known to be influential in performing cures upon grown persons.

In the application of animal magnetism, the general rule is that the person who imparts the influence should be healthy and vigorous; and it has been observed that when employed by aged persons, whose physical powers are declining, it is less powerful than when exercised by those of middle age. But in spirit magnetism the case is different. Here it matters little whether the operator is old or young, strong and vigorous in physical constitution, or comparatively feeble and delicate. The curative element is not, in this case, derived from the material body, but is imparted through the operator's organism, as the channel of transmission. Some of the best healers are far from being large, robust or entirely healthy in themselves. Persons in the animal plane of life's development, although apparently abounding in physical force, have less power of endurance in exercising the healing process, than those much feebler than themselves, who possess however the potent quality of spirit magnetism. In the former, although magnetic, the power is derived mostly from the human body; in the latter it is received by and imparted through the organism of the magnetizer: it is an invisible, subtle power, without which but little good can be accomplished in the way of curing the sick.

Notwithstanding the decided advantage possessed by the physically feebler operators mentioned, by reason of their defective physical condition being more than compensated for by their receptivity, it seems to us that the best condition for a magnetizer is to be physically strong, but with the material body *spiritualized*; and we do not doubt that all who have fully investigated the subject will have arrived at the same conclusion.

We make no pretension to the attainment of all that is to be learned concerning the operation of the law which governs the invisible power; nor will any thinking, reasonable person expect us to be able to explain unerringly why these things are so. As well might it be expected that the finite mind should be able to fully comprehend the infinite. Each individual will understand the law more or less thoroughly, according to the extent of his capacity and interior unfoldment. Some get clearer views than others, by constant study and observation, yet there can be found "none perfect." All should try to learn as much of life and its laws as possible, that they may be enabled to avoid the infraction of any natural law of their being, physical, moral or spiritual; and escape the penalties which inevitably follow transgression.

The operation of the curative process we consider to be a *chemical* one, carried on in a natural manner; that the cause of disease consists of a want of *vital force*, and an inequality of the circulating fluids; and that if a given case of disease be curable at all, whether it be internal or external, acute or chronic, it is eradicated simply by assisting *nature*, by a process which increases the vitality, thereby giving new life and vigor to the entire system of the patient. If a patient is cold, and the bodily functions inactive, magnetism is required; on the other hand, if there is fever, and the patient complains of a sensation of burning heat, electric force is needed. In both conditions it is the surplus of one or other of these forces that creates disease, in many instances; and the vital forces are brought into a state of equilibrium during the chemical change that constitutes the cure.

The magnetic and electric forces, so subtle in their nature, and almost magical in their effects, differing as we have said in quality, must differ also in their adaptation to the variety of cases to be treated, and it is necessary therefore to be guided by some practical rule in resorting to their use. But as healers differ in the degree of power which is exercised through them, and in the facility with which they severally employ it, no rule can be adopted as to the number of passes necessary to be made, nor the length of time to be employed in each magnetic operation: such a rule cannot be fixed and uniform. We have known a magnetizer to produce an entire change in the condition of his patient by a simple touch, showing that when there is a surplus of either force, its relative proportion can be changed, by bringing into operation an opposite element, adapted to the needs of the case.

Some healers possess much of the soothing element, but have not the healing power in any great degree; consequently they cannot arouse the dormant energies to activity, which is often necessary in chronic disease. Others are so fortunate as to possess both these qualities combined. They can impart the precise force in quality and degree, that is required to bring about an equilibrium in any given curable case. Hence they can reach a larger number of cases of different organizations; and their success is greater than follows the treatment of those whose ability is limited to the employment of but one of these opposite forces. Some are on the material or animal plane of life; others are on the higher or more spiritual plane of development or growth, which elevates their character, and enlarges their usefulness. Patients in selecting the force to be employed, should not confound the soothing influence with the more potent energy. The soothing process is more agreeable, but less efficacious; and in certain morbid conditions it will be powerless for good. What is needed before such persons can rely upon themselves, or inspire their patients with confidence, is further unfoldment, by which the adaptation of quality of force will be imparted in each particular case, at the time of treatment.

One magnetizer should not be exchanged for another

without proper cause. If the condition of the patient is steadily improving, it is sufficient evidence of the adaptability of the kind of force that is being employed. Such changes, after harmony had been established between operator and patient, have been known to prove fatal. And even when the result is less serious, it is but experimenting, as when changing medicines without obvious necessity. The parties most concerned should be contented with reasonable improvement, and "let well enough alone."

The habits of the magnetizer must be in accordance with hygienic laws; otherwise his usefulness will be impaired. Some think that when exhausted by over-exertion, they must recuperate their strength by a resort to the use of stimulants, which when taken to excess intoxicate. But this practice has a tendency to draw towards them a low class of influences. The practice becomes a habit; the system as it becomes accustomed to the stimulus, gradually requiring increased quantities daily to satisfy the demand. Partaking of their favorite beverage more and more freely, they soon lose self-control, and at last become a disgrace to the noble work in which they had been engaged. "The most proper course to pursue is to live a natural, harmonious life. Let the surface be as smooth as possible, and let there be as much harmony as possible between the internal and the external. Seek for that holy peace which no circumstances of earth can infringe upon. Do unto all others as you would that all others should do unto you. Live naturally; live temperately; abstain from all the excesses of life, and seek to bring all the faculties of your body into proper exercise, forgetting none. Do not exercise one faculty to the detriment of any other, but seek to exercise all harmoniously, so that the subtle currents upon which the spirit power is dependent may not be obstructed; in a word, live natural and harmonious lives. Do this, and if there is any latent mediumistic power within you it will be sure to come to the surface."

The healer should appreciate the nature of his vocation, and when it has been proved that he is endowed with the invaluable gift, he should be imbued with confidence in its power and usefulnsss. "For in order to study and practice medicine well," says a wise man of our time, "we must give it great importance, and in order to do this we must believe in it."

Says another writer, "I know of no more revolting position for a conscientious man, nor a more ridiculous one, than that of a physician who has no confidence in the means he employs. Such a man cannot possibly pursue the study and the practice of his art with the zeal, application and assiduity which alone can secure success."

Healers should be governed by principle: they should be firm and decided in character; as true to the duties of their calling as steel is to the magnet. Reason should always be at the helm. They should neither be bought nor sold; nor yield to any temptation which seeks to allure them from the path of rectitude. There are those who possess the healing power in a considerable degree, whose mental faculties are not evenly balanced; who say and do many things that are objectionable. This is to be regretted, though there seems to be no way to prevent it. Yet neither injudiciousness of speech nor erroneous conduct invalidates their capability; if well directed their usefulness will be demonstrated. The magnetic healing power is still there; and as a mode of practice it should not be condemned because of want of consistency in the conduct of some of its practitioners. Their unfoldment is imperfect, hence their practice cannot reach the highest state of perfection. Notwithstanding this, we must at last reach the conclusion that the law of magnetic force exists in nature, and that those who are susceptible to its influence will be attracted by it whether they are well balanced or not.

While these are to a certain extent successful in the use of their gift, the untoward circumstances alluded to complicate it, and impair its usefulness. In comparison with such, those who live a well ordered life have greatly the advantage. It is a fact readily deduced from observation that the most evenly balanced individuals, physically, mentally and morally, and those who are also spiritually unfolded, will be the most successful in effecting cures. There is a moral effect in being exemplary in one's daily walk, which quiets nervous irritation, and inspires confidence in the mind of the patient, and increases the susceptibility to magnetic impression. This method of cure can be made as practical in its application as any other, while it is more natural, and consistent with scientific teachings.

Many healers effect cures by the simple process

called "laying on of hands," without a word being spoken; others employ language at the same time, speaking in the most positive terms, and in an imperative tone. The latter exert a strong psychological power. Many cure by entering the room occupied by the sick person, or by sending some material that has been magnetized when absent and at a distance, as a piece of cloth or paper, or a letter addressed to the patient. By some subtle, and to the astonished inquirer, incomprehensible process, the magnetic influence is conveyed by the article thus permeated with its occult quality. And again in other instances, even this medium of communication has been dispensed with. The *mind* of the magnetizer being active, the will power has been exerted without the intervention of physical substances, and in many cases with decidedly beneficial results; the subtle vital force being transmitted directly to the patient, who receives the benefit, while unconscious of any active agency being employed for the purpose. Some of the best cures that have been wrought through our own organization have been accomplished during a single interview, and without thought on our part as to whether the patient had been benefited or not. Some magnetizers heal through the natural law of sympathy, taking upon themselves the symptoms of the patient and the desire for relief, and by making passes upon themselves, appropriating as it were in imagination his condition, relieve the symptoms of those with whom they are thus singularly brought in sympathy.

The following remarkable case in point occurred in Kentucky several months ago, and excited considerable newspaper comment at the time.

ANIMAL MAGNETISM IN THE COURTS.—A very remarkable case has recently been before a jury in Louisville. Robert Sadler, about forty five years old, an Englishman by birth, was arraigned on a writ of *lunatico inquirendo*. Those bringing suit admit that he is perfectly sane on all subjects save one, and that is animal magnetism or mesmerism, and on that he entertains the most radical and extravagant ideas. Although the suit appears not to have been brought in a spirit of malice, Sadler alleges that it was instigated by a desire to deprive him of an inheritance that falls to him from his parents in England, who are very wealthy. The form taken by the so-called lunacy is that he claims to suffer the most agonizing pains, in sympathy with people undergoing amputation, fractures, or other torture, and that, too, when he could have no knowledge that operations or accidents have been undergone. He says, for instance, than when any one with whom he is connected by this magnetism is hurt, he can feel the pain, though he does not know that it is acute. He once felt the burning process of a knife cutting in his side, and experienced inexpressible agonies, and afterwards learned that such an operation had been performed the same day at the hospital in Louisville. Other examples are given as follows: once he was sitting with his family, on a quiet Sabbath afternoon, when he was seized by a crushing sensation in his shoulder, and imagined the crushed bones distinctly in his shoulder and arm, and suffered the consequent pain, though nothing apparent to the sight was the matter with the limb, and next day the city papers brought him the news of a man being thrown under the wheels of an omnibus and having his shoulder and arm broken. At another time, while an operation on the throat of a person was being performed in the hospital, he felt distinctly the operation being made and the tube being thrust into his throat. This continued three or four days, and he afterwards learned of a similar operation being actually performed. One evening about sunset he felt a bullet strike him in the forehead, and though he was not shocked or knocked down, suffered as much pain. In five minutes after, his little nephew came running to the house, saying that a man had been shot a short distance from his house. He declared furthermore that he would scream in the night from pain inflicted upon others, and gave further instances of sympathetic suffering. His testimony was perfectly rational, and the jury being unable to agree were discharged. This seems to be a case for serious scientific investigation, as the man's character appears to be above suspicion, and it is positively stated that the coincidence of suffering was in many cases clearly proved.

This candid statement is taken from the Newark Advertiser. The case was subsequently submitted to the controlling mind of the Banner of Light circle for explanation, and the following is the response. "The time has gone by when the cry of "hallucination," "deception," can be successfully raised with regard to such phenomena; for as men and women are beginning to understand more and more clearly concerning the science of life, these absurd ideas are beginning to be more and more clearly understood by them. You talk of space, but in reality there is no space. The atmosphere by which you are surrounded is filled with innumerable threads, magnetic and electric, binding all souls together and all bodies together, and it is by no means outside of the science of life that we find somebody so exceedingly sensitive to the action of other bodies as to be able to feel, to sense all that is transpiring with those other bodies with which they are in magnetic and electric *rapport*. It is a well known fact that the mesmerizer transmits to his subject feelings of pain or pleasure, sorrow or joy; anything, either mental or physical, that may be an experience of his own, is also made an experience of the subject. This has been demonstrated again and again, till it has become a settled scientific fact, so it is very easy to see that the phenomenon in question is of the same class, belongs to the same head. It is nothing more miraculous, nothing outside of the order of Nature, nothing beyond the science of life, but within the range of natural, simple law."

A person in the audience stated that he supposed he had experienced a similar phenomenon in connection with the person of a friend who had died of paralysis, which continued for hours after the decease; and inquired whether it was not the effect of the diseased magnetism upon the nervous system. The reply was, "It was doubtless the effect of what you may be pleased to term the diseased magnetism of the patient. It was transmitted to you not simply through the law of sympathy, but through a more potent law, and one that acts whether you will or no. It was transmitted, doubtless, by your being in electric *rapport* with the individual. There is a difference between being in electric and magnetic *rapport*. One belongs to the positive, and the other to the negative. That which comes to you through the negative, comes to you without any consciousness on your part. That which comes through the positive, comes through a consciousness on your part. But they are equally potent, and belong to the same general law of action and re-action that is being manifested throughout all forms which have an existence."

Of the connection of mind with electrical and magnetic influence it is observed in a discourse on the Origin of Mind, that "All mind when it proceeds from God is harmonious. It is apparently inharmonious when it comes in contact with inharmonious substances—hence it appears to move as though inharmonious in itself. This accounts for the inharmonious effect produced upon material substances—bodies under the operation of Spirit influence.

The inharmony between the two substances, the animal-electricity and the spirit-electricity, produces various physical demonstrations and conflicting disturbances. As the material and spiritual become more assimilated, the effect is harmonious, producing a most delightful soothing sensation or influence. This influence mankind have named Magnetism or Electricity. It is effected either by direct or indirect spirit influence

—by direct spirit influence when the spirit power acts directly without the assistance of other than its own power, by indirect spirit influence when it acts through another mind and body as a medium. This influence may be greatly retarded or accelerated by the condition of the mind and the thought of the operating medium. Hence the advantage of natural mediumship over all artificial methods of developing processes. No mind is in itself independent; and while it is dependent upon and influenced by other minds, it is liable to imbibe the particular views, prejudices and sentiments of those minds, increasing its own resistance to pure spirit-truth, if those views do not harmonize with truth."

We learn also that "The mind does not decay. Its manifestations become imperfect, because here in mortal life it is called upon to manifest through a mortal machine; and if that is out of order, if that has become diseased, the manifestation will be correspondingly diseased and out of order. The most perfect musician cannot give a perfect manifestation in music unless you supply a perfect instrument;" and that "human beings in proportion as they are purified and regenerated, or truly spiritualized, become receptacles and channels for spirit magnetism; or to use other terms, become Leyden jars and batteries for accomplishing the potential personal force and distributing it to others. Hence the propriety of the anciently originated custom of imparting the Holy Spirit by the laying on of hands."

It has been contended that the power of restoring the sick to health by the means herein described, was a special dispensation confined to a few individuals at two different epochs many centuries ago. But if such things were done in olden time, why cannot they be done in this age? The human constitution physical and mental, has undergone no material change during the time that has intervened. Nervous impressibility is the same, and the ills that flesh is heir to, have similar needs. Jesus said that not only the same, but greater deeds than he had done, should be performed. And indeed the case is not dependent upon the argument founded upon the similarity of relative circumstances since prevailing, nor upon the prophecy, for the facts of subsequent history prove that the healing power continues in existence; that it has descended to the present generation, and is now exercised by large numbers of persons in our midst, with distinguished success.

MORAL INTEGRITY OF HEALERS ESSENTIAL.

THE consciousness of a just cause which springs from a faith in its truth and goodness, increases the degree of success obtained in its service. One of our most respected healers remarks "I am a firm believer in spiritual influences—no man can be a good healer without it." Would that all who are engaged in this field of labor were thus convinced. Some who are called to it have but an imperfect understanding of its nature and duties. They know that the source of the power exerted is beyond themselves, and that it is exerted independent of, and sometimes in opposition to their will. They are honest in their purpose, and require only to study their gift, to exercise it with satisfaction to their own minds, and with benefit to the afflicted.

There is another class who are dishonest and mercenary. They infest every city, town and village, and are distinguished for making the "almighty dollar" their first and often exclusive consideration. "By their fruits ye shall know them." Persons of this class possess but little healing power, but have an abundance of self-assurance. They advertise extensively, and by their assumption force the timid and credulous to risk their treatment. By the impulse which their self-assurance gives the patient, they occasionally do a little good. Their next step is to procure a certificate to the alleged fact that a cure was effected at the first visit, or consultation. Sometimes they bribe the patient to write it out, and where this cannot readily be done, they do not scruple to write it themselves. By such stratagem a certificate of cure is sent abroad before it has been ascertained whether the patient has really been even partially benefited. Not long since we met with a case of this kind. The patient died, and for months after the fatal termination of the case, the "healer" was distributing his circulars broadcast over the country, with the certificate of the "remarkable cure" blazoned upon it! When the relatives of the deceased patient were made aware of it, they expressed their indignation at the course pursued by the "healer"; they published a refutation of his false statement in the newspapers, and threatened him with a prosecution if he did not desist from further deception. This person advertised as all healers do; but he did not acknowledge receiving aid from an invisible source, but claimed to accomplish his feats by his own personal skill. Sailing under false colors, those who had not made themselves acquainted with the phenomena, were unable to detect the difference between his pretensions, and the genuine acquired ability of a conscientious magnetizer. Hence deception with him was an easy task.

There is another class who employ persons to present themselves at places where there are large numbers of people assembled together, on pretence of being blind, lame or deaf; and the pretended healer restores them before the multitude. The witnesses of these "miraculous cures" observing the apparent evidence of a healing power, not examining critically, take it for granted that the alleged cures have been actually performed, and so report. If they exercised ordinary prudence and watchfulness, they would detect the unreliability of the pretender and his accomplices, and arrest his nefarious operations. As it is, after their open endorsement, many flock to see the "great healer," who quickly relieves them of their money, and takes his departure in search of new dupes. The patients, experiencing no other *relief*, and being unacquainted with the practice, naturally infer that it is a failure in the magnetic method. They take it for granted that all healers are alike, and that when a pretender fails, a genuine healer must also fail; and then set down the practice as a "humbug" without stopping to investigate the operaton of the law, in accordance with which genuine cures are effected. When people do not think for themselves, but exercise blind credulity, it is not surprising that they should remain ignorant of that law, and become the dupes of the designing. Let them investigate this subject for themselves, as they would any other, with equal deliberation and candor, and they will soon learn how to keep out of the hands of unprincipled adventurers.

These remarks apply with equal force to the vending of quack nostrums or patent medicines. The itinerant vendor collects a promiscuous crowd of persons about him, and by his bombastic pretensions induces the credulous to try his wares. Where pain exists, the pungency of his nostrum will sometimes momentarily blunt the sensibility, or divert the attention. Cures are supposed to be made, and the patent medicine becomes renowned. The effect is but momentary; the cause still remains, and the pain returns with renewed severity. True healing is effected in accordance with one of the fixed natural laws, as uniform and harmonious in its operation as any other in the Universe of matter and mind.

The public mind should be better informed upon this mode of practice: therefore it is that we state facts that have come under our own observation, and for the benefit of the practitioners of magnetism, and of all who are interested in the promulgation of its truths. With an intelligent writer we may exclaim, "May God and the angels help us to redeem the divine art of healing from all imposture and quackery!"

UNCONSCIOUS MAGNETISM—SLEEP —THE MARRIAGE RELATION.

BESIDES the employment of magnetism with a view to its curative power, there are numerous examples of its operating imperceptibly, and without design or motive on the part of him who imparts it. Such persons in the enjoyment of good health, affect the sick favorably or otherwise whenever they come in contact with them, or within their sphere of attraction; and doubtless it operates much more extensively when not recognized as a power, or its effects not understood. It may then be termed antagonistic magnetism.

These results, brought about imperceptibly, are perceived after a time when persons of different ages and conditions sleep together. There is the foundation of a great amount of suffering laid during sleep, by the emanations from persons who are antagonistic in their chemical magnetic forces to others with whom they sleep. Those who thus unconsciously radiate an injurious magnetism from their bodies, may be harmonious in spirit, and correct in their intentions and moral deportment, but the magnetism which is the life force of the one may, for the time being, be repulsive, injurious, and to a certain extent destructive to the other; a change constantly taking place in the relative condition of both. In such cases the remedy consists, of course, in directing the parties to sleep apart. With an intelligent appreciation of the necessity of temporary isolation, giving time to regulate and harmonize the opposing forces, the remedy can be applied without any misunderstanding, or unpleasant disturbance of the moral relations.

In a publication entitled the *The Laws of Life*, the following statement is made on this subject, "More quarrels arise between brothers, between sisters, between hired girls, between school girls, between clerks in stores, between apprentices, between hired men, between husbands and wives, owing to electrical changes, through which their nervous systems go by lodging together night after night under the same bed-clothes, than by any other disturbing cause. There is nothing that will derange the nervous system of a person who is eliminate in nervous force like lying all night in bed with another person who is almost absorbent in nervous force. The absorber will go to sleep and rest all night, while the eliminator will be tossing and tumbling, restless and nervous, and wake up in the morning fretful, peevish and discouraged. No two persons, no matter who they are, should habitually sleep together. One will thrive and one will lose. This is the law, and in married life it is defied almost universally."

Aged persons receive benefit from sleeping with the young; the young person is injured by the magnetism of the old. Some persons are so constituted that their emanations during sleep are productive of mutual benefit. In other instances one is benefited and the other injured. The law regulating this influence holds good with relations and intimate friends, and belongs indeed to the whole human family. Its remarkable effects admonish all to learn the laws of life; and when it is once known that under certain circumstances deleterious results are produced, and the remedy therefor is clearly pointed out, the exercise of judgment and reason will dictate a resort to the remedy without hesitation. Harmony will be brought forth from discordant elements, sickness will be less frequent, and the solid pleasures of life will be greatly enhanced.

"Scientific men have determined that it is best to repose with the head towards the north. There are two distinctive poles to the brain, north and south; and it is presumed by placing the head northward during the time of repose, that the natural functions or animal forces recuperate much faster than when placed in any other position, from the fact that the position is to the physical body the most natural one. Whatever course is most natural, you move in with the least inharmony. If you in ignorance trample upon the laws of your nature, inharmony, warfare, is inevitably the result. But when you place yourselves in harmony with the laws of nature, harmony and peace is the result.

"It is believed by many scientific men that did the

human race understand themselves and their connection with natural laws by which they are governed, they would enjoy almost perfect health. Disease can only be the result of ignorance, which is but imperfect growth. As you grow into a knowledge of yourselves and the laws governing the phenomena by which you are surrounded, you will grow out of a condition of disease into a condition of health. But as all progress moves by slow processes, so this thing will be accomplished very slowly; and thousands of years must elapse ere disease shall be swept away from the face of the earth, and love and wisdom become twain in the flesh."

The elimination of magnetic influence from the organization of the person imparting the force taking place involuntarily, unconsciously, and in obedience to natural law, although antagonistic to certain others, and detrimental, such person cannot be held responsible for the results, until he has had a fair opportunity to learn the operation of the law, and its consequences. Then with knowledge comes responsibility. He cannot thereafter disregard its teachings without laying himself open to condemnation. The more negative a person is, the more susceptible he is to magnetic influence; and his over-sensitiveness should entitle him to commisseration, and he should be instructed how to avoid the undue influence of antagonistic magnetism.

Clairvoyants have often described the emanations that are seen around different magnetizers, as well as those seen radiating from persons who do not possess the power in any marked degree. According to information derived from this source, there is a difference in the color of the particles thrown forth. It is reasonable to infer from this circumstance, even if direct proof were wanting, that there is a difference also in quality, as it is known to differ in quantity. If every one had the faculty to perceive this, we should doubtless see the rays of the electric and magnetic forces shooting from every part of our persons, more particularly from the hands and head. This is depicted in the halo surrounding the heads in some of the works of the old painters, who seemed to have an intuitive perception of the pheomenon of radiation, and a correct understanding of its nature. And it is not difficult now for those of us who are not blessed with the "gift of seeing" to imagine and to form a true conception of the fact which others more definitely realize. Neither can it be considered strange or mysterious, in view of the difference in quality and quantity of eliminated force, that persons can be cured of disease by a simple touch, or by transmitting the same force through material substances, as in the case of letters and other articles heretofore mentioned, to distant points. If this can be done by virtue of an imponderable agency, as stated, there can be no doubt of its efficiency when projected without the intervention of material substances.

What has been denominated by Baron Reichenbach *od force* is in all probability identical with those under consideration. Said one of the higher intelligences in reference to it, "It is a force which he himself understands very poorly. He calls it thus, because he does not know how to analyze it—because it seems to be distinct from animal and spiritual magnetism. But he is mistaken. If it is not a part of the great magnetic life by which the forms existing upon the face of the earth are sustained, I am sure I am at a loss to determine what it is."

The marriage relation affords an illustration of the law of unavoidable spontaneous attraction and repulsion. Much inharmony exists between some who have previously lived together harmoniously in that relation. They have become discordant, and are living wretched lives; so much so that they consider it necessary to invoke the aid of the statute of divorce, to dissolve the once silken bonds; or look for a release when nature's law of dissolution shall have dragged its weary length along.

In the majority of such cases, where there was harmony originally, and neither party had been subjected to the psychological influence of the other, we believe the cause of unhappiness to depend simply upon a chemical change taking place in the condition of the life forces of the parties in antagonism. What was once attractive has become repulsive. In such cases it requires but a short time to pass through the change; and if the parties will conduct themselves as becomes reasoning and morally responsible beings, exercising mutual forbearance, agreeing for the time being to disagree in matters of minor importance, they will become harmonized, and will be as likely to live together thereafter in the enjoyment of mutual esteem and affection, as they had ever done before, or as they can expect to do by contracting new ties.

Alleged incompatibitity is not irremediable. It is not only a moral but a physiological question; and it is proper in this connection to allude to a term which in consequence of its perverted use has become a by-word. We refer to the word *affinity*. In the common acceptation the terms "attraction" and "affinity" are nearly synonymous, and are employed alike in natural philosophy and chemistry, metaphysics and morals. It is a peculiar power which when operating under law, brings together elementary particles and chemical compounds from natural selection, and by moral choice unites mind with mind. It explains certain phenomena in the animal, vegetable and mineral kingdoms, as well as certain fixed laws of spiritual life. Affinity in its perverted moral sense embodies a dangerous doctrine. It is made to convey the meaning of a license to selfish sexual indulgence, as "free love" is made to mean "free lust," which latter immorality is injurious to both physical and spiritual well-being. Persons who have yielded to the seductive influence of affinity in its perverted sense, dissolving the sacred bond of conjugal union, without adequate occasion, and have formed new alliances, it is found after a time, were merely capricious and discontented; and they are no happier in the new relation than they were in the old.

Where inharmony has manifested itself, let the apparent cause be referred to whichever side it may, or be the result of mutual misunderstanding, the case should be considered a pathological one, and the parties treated as if afflicted with a nervous affection; then will the progress of antagonism be arrested. In a large majority of cases, the temporary evil, if the parties are honest and sincere in their desire to do what is just and right, will regulate itself. They should wait patiently for the end of the unseen chemical change of the life forces that is going on. The justice of equal rights should be recognized and submitted to by both. Husbands should not ask of their wives that which they themselves would be unwilling to grant. Each should bear an equal share of the burdens and responsibilities of life, equally performing its duties and sharing its joys.

A correct understanding of the subject would greatly diminish domestic infelicity, and divorce would never be required except in cases where a great wrong had been suffered by one of the parties, who had been induced to enter the marriage state under psychological, or undue and improper influence. These are very soon understood in all their disgusting realities. The one party has been actuated by a selfish, lecherous and evil motive; the other has been confiding, unsuspicious, seemingly in a semi-trance condition, scarcely responsible for moral action. As soon as the influence has lost its novel power, the charm is broken; then comes discord and reproach. The mischief is irreparable; and even repentance and reformation on the part of the evil-doer in such a case can hardly bring restitution to the injured.

DETRIMENTAL INFLUENCE—INSANITY—OBSESSION.

Many negative persons are affected by controlling malevolent influences in their ordinary intercourse with society, which it is almost impossible for them to throw off; and after being brought under such an influence, the sensitive recipient will be very liable to an attack of severe illness, unless access can be had to some good magnetizer, or sympathizing friend capable of neutralizing the force and averting the consequences.

Persons when acting in the capacity of nurses, whether by day or night, or visiting a sick friend, when in a sympathizing frame of mind, are liable to be afflicted in the same manner as the sick person. Magnetism adapted to their sensitive condition, will soon bring about an equilibrium, and restore them to their usual health.

A state of exhaustion arising from over-tasking the mind or following excessive bodily labor, increases the tendency to be affected by diseased persons, and to *take on* their condition; as well as rendering them more susceptible to malignant influences, than when in an equable, healthy and positive condition. Persons affected in this way have become excited and committed extravagant acts, which has resulted in their being placed in insane asylums, who might have been restored by magnetic treatment in a very short time.

When persons become discouraged from failure of their plans in life, or affliction or mental trouble of any kind befalls them, they are easily affected by this class of undeveloped influences. Those who are not fully aware of this increased susceptibility, and the unhappy results of the bad influences which surround us all, and by which we are liable to be painfully affected, when in an inharmonious condition, will pronounce the disordered state a case of insanity. In such cases, all that is necessary is to find a magnetizer, whose power is adapted to the disturbed elemental forces, place the person in favorable surroundings, and he will be restored as readily as if afflicted with any other disease. Many cases could be cited to prove the assertion.

The over-tasking of the mental powers is too serious a matter to be lightly set aside. It has recently arrested the attention of scholars and physicians in our own city, in consequence of the forcing process too common in the higher grades of schools. An important petition was presented to the School Committee asking that saturday may be made a full holiday in the Latin School. The signatures amounted to four hundred and twenty one, nearly one half of whom were parents of scholars who were pursuing their course of study in the school. One hundred and fifty three were physicians, and of these some took occasion to express, in an especial manner, their opinion of the necessity of a more moderate plan of instruction. One physician remarked, "I have two boys of fit age to send there, but decided that I could never do so under the present arrangement." More startlingly the Superintendent of the Boston Lunatic Hospital says, "I cannot doubt that our modern system of forcing the tender brain of youth lays the foundation for the brain and nervous disorders of after years—the causes of melancholia, paralysis, softening of the brain and kindred diseases, becoming so fearfully prevalent." Others relate the evil effects which they have experienced in their own families, seeing their sons broken down in bodily or mental health, as the result of the over-straining system.

"It is a well known scientific fact that when one organ is diseased and incapable of performing its functions properly, some other organ must bear the burden of labor—do what the diseased organ fails to do. For instance, if the lungs are weak and cannot perform their function, the labor is thrown upon the liver, the brain and the heart—generally upon the heart, and it must labor all the harder because the lungs do not work. Now, in cases of lung difficulty, when the upper portion of the lungs is diseased, the brain is found to be exceedingly active. Hope is always very active. So the consumptive is always hoping to get well; seldom ever thinks that his disease will prove fatal. The brain, under certain physical ailments, becomes very active, and because of its activity, the spirit manifests itself clearly, is not deficient in power, because the brain is not diseased. It is performing its functions well, and more than well, doing the labor for the lungs and other weak portions of the body that under ordinary circumstances it would not be called upon to do."

With reference to insanity we may again quote a paragraph or two from the same source of intelligence heretofore alluded to. In answer to an interrogatory concerning the possibility of preventing an outbreak of the disease by education and self-discipline, in those who were aware that some of their immediate ancestors had been insane, it was replied as follows:

"Medical men tell us that it is almost impossible to prevent hereditary insanity; that is to say, unless you know just where to strike, you are very apt to strike in the wrong place. Now, as insanity, as I have remarked, is located upon and through the imponderable forces, it is a more subtle disease, and does not become apparent often until it suddenly bursts upon you in all its fury. Medical men tell us that the seeds of insanity are very frequently sown at conception. Then it is called hereditary. It is transmitted from the ancestors down through a direct magnetic and electric line. If you know that your ancestors have been thus afflicted, the only proper and sure course is, if you wish to stay its progress, to avoid marriage. Medical men tell us that when once the disturbances are in the imponderables of the body, you can rarely affect them for good, except at the time when they have shown themselves the most violently—when they have reached a certain point, then you are able to affect them (if you know how to apply the agents,) generally very successfully. But even if you know that you have the seeds of insanity implanted within your being, you can do nothing towards eradicating them till they have shown themselves outwardly. Now this seems rather hard, but those who seem to understand such things declare that it is absolutely true."

In reference to climatic influence it was stated that as is well known, it was very extensive, but that the climate which would be most favorable to one individual may be the most unfavorable to another; therefore no rule could be laid down with reference to it, which would be applicable to all.

The application of magnetism as a curative means in connection with the treatment of insanity, is given in the following words. "Medical men inform us that insanity is simply an unbalancing of the physical and spiritual forces. They inform us that the cause is seldom found in the physical organism alone; but it is found with the forces that play upon the organs. Therefore it is very hard to know exactly how to treat the different kinds of insanity. They tell us it is a very subtle disease, sometimes appearing to yield to remedial agents, and suddenly rising up again with more vigor than ever before. Medical men—in the spirit world, not here—inform us that they are doing all it is possible for them to do toward enforcing their ideas of insanity upon the plastic brains of medical men on the earth. Those who are the most susceptible to spirit influence will receive their ideas first. I believe that the foundation of their theory is here: Insanity lying in the imponderable forces, should be treated not as you would treat organic disease, but as you would treat spiritual disease, or a disease running through the imponderable forces of the human body. Magnetism and electricity have been heretofore very little understood. They have been recognized as existences, but their wondrous uses have never been sought out. Now medical men inform us that magnetism and electricity are the most powerful agents that can be used, if used understandingly, in all cases of insanity, but inasmuch as medical men have so small an understanding concerning these forces, it would not be safe for them to seek to make use of them till they have learned something more of them. Magnetism and electricity stand as masters over humanity, but when humanity comes to know these agents humanity will master them, bring out all their uses and apply them to the needs of the suffering."

We are confident that many of the cases that are termed insane, can be restored in a short time by the magnetic method of treatment, even when considered incurable by ordinary means. This is so apparent to us that we believe it will not be many years before magnetism will be one of the principal means of treatment adopted in all hospitals, that capable magnetic physicians, both male and female, will be recognized by the constituted authorities, and that the great value of the practice will have been so fully demonstrated, that a resort to it will be advised by medical practitioners generally, in cases where medication has failed to produce any beneficial result. There should be institutions established in all sections of the country, for this class of sufferers which shall embrace this as one of the most essential means of restoration to mental health.

We come now to that singular phase of human experience known to investigators of spiritual life, by the term obsession. This form of disease is almost unknown to medical practitioners, because but few of them have studied its phenomena, and consequently the mass of them do not recognize its existence. It is frequently taken for insanity. Its outward manifestation resembles it, the symptoms being in most particulars the same. Medication affords no relief to the afflicted one. Sensitive persons are frequently affected in this way, when unconscious of any influence being at work beyond the range of their own personal responsibility. Many get so completely controlled by the invisible influence, that during the paroxysms their entire character is changed. Physicians despair of curing them, but consider it necessary that they should be placed in an insane asylum, from a regard for the safety of the sufferer, as well as that of the public.

As these cases are becoming quite common in the present age; as much so as in olden times, it is important to know what course to pursue, when a person not well versed in the phenomena, and particularly the true cause, is the subject of attack. It is useless for those who dispute the fact of obsession to rest their case upon simple denial, because they are not satisfied with the alleged cause. It is a natural occurrence; as plainly so as when persons suffered the same afflictive experience in past ages. The cry of "humbug" does not settle the question. It does not set aside the facts, nor relieve the suffering.

It is the duty of all persons to learn the true nature of this unhappy affliction—indeed whether, as we claim, such a diseased condition as obsession has any real existence; both for their own benefit, and that justice may be done to the sufferer who would otherwise be blamed for conduct which he cannot avoid; and that a rational and humane system of treatment may be adopted. The affliction from whatever cause it may arise, is liable to befall any one: neither Jew nor Gentile, Christian nor Infidel can claim exemption from it. Patients are as easily relieved from it in this day as they were in the days of Jesus and his apostles. Most of the diseases cured by them in those days arose from this cause.

History abounds with cases illustrating the manner in which these unfortunate victims were controlled by undeveloped spirits, the young, the pure and the good being alike subjected to their baleful influence. Even little children whose nervous organisms rendered them susceptible, were taken advantage of. But singularly nearly all the notorious cases of such obsession were those of persons who were not adherents of any system of spiritual philosophy, and consequently were considered the most unlikely subjects for its exercise. The Salem Witchcraft commenced in the family of the Rev. Mr. Parris, seizing upon his innocent children, and causing them to perform revolting actions, similar to those of animals and even reptiles.

The Witches and Warlocks mentioned in Scottish history, whose remarkable doings took place in the middle ages, the account of the Nuns of London, the Convulsionaires of St. Medard, the Tarantula dancers, and the religious Monomania, as it was called, in Sweden, are all examples of obsession; and they are all epidemic outbreaks which according to religious sectarian preconceptions, affected persons the least likely to have been excited by enthusiasm or fanaticism. A thrilling account of similar occurrences which took place in the valley of Morzine in Switzerland, beginning in 1857, is given in the *Cornhill Magazine*, published in London.

The citation of a few cases will be sufficient to show the characteristics of this malady. The first is from an official report by Dr. Constans. "The patient was about thirty years old. She was married, and the mother of a family. She was dark in complexion, and of a nervous temperament: her health was good. At the time of my visit she was making preparations for going to Salenches, a town at some distance, where she was to be sequestered. When I saw her in her room leaning over her baggage, I spoke to her, but she did not reply; soon after her head and upper members became convulsed, and she began to speak in a jerking way. I pinched and pricked her unawares with a large needle, as she leaned against the table, but she gave no sign of pain. Presently she threw herself on the ground, and rolled about and struck at the furniture and floor with extraordinary violence. Her face was red; her throat swelled; she seemed suffocated. I tried again if she were sensitive to pain, but with the same result as before. She continued to struggle and cry out.

"I am from Abondance," (a neighboring parish,) said the devil by her mouth. "I was cast into eternal fire for eating meat on a friday. Yes, I am damned," he continued. "*Mortuus est damnatus.* I must torment the woman—I must drag her with me." Then, leaping up, with one bound, the woman, or rather the devil, cried out, "I died by drowning; the woman must die that way." She rushed out to throw herself into the river, where once before she had nearly succeeded in destroying herself. Three strong men could hardly hold her back, though in her struggles she seemed to avoid hurting them. At last she desisted, and leaning against the table, she recommenced her abuse. "Ah! bearded wretch of a doctor," she said, "you want to drive us out of the woman; we fear not your medicines. Come, we defy you. See you, wicked unbeliever, what is wanted are prayers, and priests, and bishops, and pious exercises. We are five in this woman. Now there are only two who speak, but it will be very different when she passes into the country where her forefathers are buried, near the church where she knelt innocent; oh! there it is that we will torment her."

The fit left her suddenly, as with the other woman I had seen, and without any pause of transition. She passed her hands through her hair, asked her husband to give her water, and drank a bowl of it. Her replies to my questions were simple and natural. She remembered nothing of what had taken place.

It is curious that every friday she went to the mayor and asked him for bacon, which she ate eagerly, and sometimes raw. Our readers will remember that the devil who possessed her had declared himself damned for having eaten meat on friday."

The following account is taken from the same magazine:

"In the spring of 1857, the village being in its usual quietude, Peronne Tavernier, a child ten years old, was engaged in eager preparation for her first communion. She was exceedingly intelligent aud sweet tempered, and a sort of favor had been made in admitting her sooner than her comrades of the same age, to the mystery of the Eucharist. Religious thoughts occupied her, she says, night and day, and she could speak of little but her joy in the prospect of the event that was at hand. One day—it was the 14th of March—as she came out of church after confession, she saw a little girl fall into the river, and felt strange fright and uneasiness at the sight. A few hours afterwards, as she sat at school, she suddenly sank down on the bench, and had to be carried home, where she remained as one dead for some hours. Three or four days later the same thing happened to her in church, and afterwards the attacks occurred frequently wherever she might be. Again, in April, as she and another child, Marie Plagnat, kept their goats on the hill-side, they were both found insensible, clasped in each other's arms. They were carried home, and after an hour Peronne awoke and asked for bread, which however she could not eat. After that the seizures became frequent, and both children were attacked five or six times a day. Symptoms that strangely impressed the by-standers began to manifest themselves. The little girls in their trance used to raise their eyes to heaven; they sometimes stretched out their hands, and appeared to receive a letter. By turns it seemed to give pleasure and to excite horror. Then they made as if they refolded the letter and returned it to the invisible messenger. On awakening they declared that they had heard from the blessed Virgin, who had shown them a beautiful paradise. When the missive, as they sometimes averred, came from hell, Peronne used to complain with terror of serpents that were twisted round her hat. Day by day the attacks became more remarkable. The children began to gesticulate, to speak incoherently, to utter oaths, and blaspheme all they had been taught to revere. Their limbs were convulsed, so that three men could not hold Peronne in her fits. In their trances they accused men in the village of having bewitched them. Among other predictions, they announced that two other girls and Peronne's father would be seized as they were, and that the latter would die."

Cases similar to these are occurring in almost every town and city in our own country. Many have come under our personal observation, which we have been able to control and restore to health by a proper use of magnetic treatment. One of these was that of a beautiful, innocent young girl of about twelve years of age, living in Chicago, a member of a Baptist sunday school. This child suddenly became passionate, obstinate and refractory. She would spit upon and strike her mother, her actions resembling the conduct of an insane woman who formerly resided in the family. The mother was so much troubled by this change of conduct that her hair turned gray within a few weeks. The minister was sent for, but his counsel could not remedy the evil. Several physicians were employed, but their skill could not reach the case. After remaining for some weeks under treatment, magnetism was suggested as a last resort, and the result of its employment was that in a few days her appearance became healthy and her conduct proper. She then attended a methodist meeting, and spoke in language far beyond her acquired ability; information was sent to the mother that her daughter was "converted." At the last account her condition was harmonized and health restored. In this case, the the girl was undoubtedly the subject of control by two different spirits, the one undeveloped, the other progressed.

Another case is that of a beautiful young lady of fifteen years of age, residing in New York, a member of the Baptist church. Her physical system was in a morbid condition, and her conduct capricious, the cause being quite inexplicable to her friends. Sometimes she would calmly rehearse the Jewish marriage ceremony; at other times she would laugh and talk in a witty strain, exciting the risibilities of her friends. In a few moments she would try to throw herself out of the window. At other times she would talk strangely about her "dead" father and uncle. The mother incurred considerable expense for medical treatment, but without benefit. She was then placed under magnetic treatment, which soon restored her to health.

Some twelve years ago when in conversation with an intelligent gentleman, a leading methodist in Philadelphia, he related an interesting experience which he had had in relation to such an influence. He said that when he went to camp meetings he generally felt impelled to be the first to speak, and that the influence under which he spoke made him feel happy, "and," said he "I could not speak ill of any one; but in a few moments afterwards I felt like kicking every one out of the tent, and I would go away from the camp meeting and *war* with that spirit for some time. It would take a whole day before I could be myself again." When in after years he came to study the nature of the influences which are thus wrought upon human beings in earth life, the cause seemed clear to his mind, and when he departed to the other life, at the ripe old age of eighty years, it was in the full belief of the fact of both good and bad obsession.

A fourth case occurred in the person of a young woman in the State of Connecticut, who had been a sufferer from domestic trouble, and was about obtaining a divorce. She was much afflicted by the control of an undeveloped being, and was impelled to make a journey from Boston to New York bare-headed, and committed other extravagant and unreasonable acts. As soon as proper conditions could be brought about, and magnetism was employed, she was restored and saved from the necessity of being sent to an insane asylum.

An unhappy instance of obsession came under our observation in Troy, N. Y., which for want of appropriate treatment terminated differently; viz. the case of an interesting girl of some twelve years of age, who would at times write, sing, and play upon the Piano in a style far beyond what her natural skill and taste, or her acquired ability had enabled her to do. These agreeable exercises were alternated with others of a character painful to witness. There were sudden outbreaks of vulgar and obscene language; and these paroxysms continued for several years, the unruly power steadily gaining the ascendency. At length in a moment of violence she was thrown upon the fire, the hair burned from her head, and her face badly scarred. She lived a most miserable life for the remainder of her days, in the care of her mother, until at the age of twenty six years she passed to the life beyond this material existence, the change bringing great relief to herself and friends.

Somewhat akin to these is the case of a lady in New York who was the victim of paroxysms of ungovernable ill temper. She became enraged with her friends, and could not be persuaded to remain at home with her husband and children. Although under medical treatment she felt no confidence in her family physician, and seemed destitute of moral power to elevate herself from her unhappy condition. She was taken to Niagara and Saratoga with the hope of calming her excitability by change of scene, but to no purpose. Magnetism was applied, which gradually equalized her nervous condition, and in a few weeks she was restored to health, and again lived in peace with her family.

In relation to repeated obsession and persistence in holding the subject under control, we quote the following pertinent remarks:

"A good spirit will not attempt to take and hold unwarrantable possession of a mediumistic organization, hence you may rest assured from what class it is that the phenomenon of obsession proceeds. Now, if the infesting spirit were not magnetically stronger than his subject, he could not maintain possession, however he might once gain temporary ascendency. The true processes of cure, therefore, are obvious and dual. First, let all possible means be taken to strengthen the health of the subjects, and render their minds positive to the control of others. Good air, good diet, change of scene, association, and constant employment, pleasant society, and cheerful occupations, are the physical means, which steadily resorted to, may alone effect a cure. If these fail, use in connection with them the aid of a strong-willed, powerful and virtuous magnetizer. Let him continue with unflinching constancy to exert his will, and add thereto magnetic passes over his subject, and we will pledge our faith and word that he will speedily dispossess the enemy, though he were the fabled Beelzebub in *propria personæ*."

There is a class of cases of a milder type which although not in any respect like obsession, are distinguished by a debilitated and otherwise disordered state of some of the functions of the mind, and may here be mentioned. The spirit or mind is the part which suffers, instead of the physical body which contains it. When separated, the physical body feels no pain; but when connected we often think that it is the material organism which experiences the sensation of suffering. Both seem to suffer, but it is the spirit which is disturbed when the house it lives in is out of repair. Therefore the spirit needs treatment as much as the material body.

There are many cases of this character. There seems to be no organic affection; but the patient appears to have lost vital force; he becomes discouraged and loses his interest in the events of life as they transpire around him. All things seem to him dark and unsatisfactory; and if he is not lifted out of this desponding condition, active physical disease will soon be developed. By magnetic treatment he obtains relief, and becomes as it were re-juvenated. Many require treatment but once, and others less impressible, begin to improve at once, yet need a few repetitions; they however, are only in an inactive condition, and as soon as the vital forces receive an impetus, they regain their usual health. Remarkable cures have been wrought by a single application of the magnetic treatment. In such a timely manner were the circumstances made favorable for its employment, and the result so effectual, as to lead to the inference that the healer had been sent to the patient by an invisible power; or vice versa, the patient to him, by some singular coincidence. We do not regard any cure as a miracle, but as the result of the operation of a fixed law of nature: the successful employment of magnetic power being no more marvellous or supernatural than similar events daily taking place among us now. So numerous have been the cures called remarkable within the last few years, that but few persons are ignorant of the fact that some such things have been done. If each one has not been a living witness, the testimony is so strong, and so seldom wanting among friends, who have themselves been the recipients of benefit, or have known of its power being exerted among their neighbors. Many within our own personal acquaintance, who were abandoned by physicians as hopeless and incurable, are now rejoicing in health that has been regained through this agency.

NOTE. A remark may here be made in concluding this part of the subject, with reference to a main cause of what is considered insanity, and its curability. We do not doubt that of all the cases considered insane, as many as one fourth arise from a psychological cause, and it may be from the influence of minds in the human form or from invisible disembodied minds. We believe also that by magnetic treatment one half of the patients now confined in asylums, could have been restored to their ordinary mental soundness in a short time, as readily as if afflicted with any other form of disease.

The concentration of thought upon any one subject for an inordinate length of time, leaving inactive those faculties of the mind not concerned in that thought, is calculated sooner or later, to destroy the equilibrium of mental action. It matters not whether the conclusions arrived at are correct or otherwise, or that the thought has been in a proper train. It is the over-straining, as in the forced educational system mentioned in the text. Especially is this dangerous when the mind is much exercised upon religious subjects, that is to say those involving doctrinal or controversial points. And these we are told by alienists, when once settled down into a diseased condition, are among the least hopeful. Speculative ideas which enter into the formation of creeds of various sects, and some which are creations of individual fancy, are widely prevalent; and this gives encouragement to the weak and superstitious to dwell upon or make hobbies of them. Hence caution should be exercised in reference to speculation upon all subjects which have a tendency to over-excite the brain and nervous system.

It has been a matter of curious inquiry, to what extent obsession prevails. This can only be satisfactorily ascertained by clairvoyance. In answer to the question recently propounded to an intelligent authority, the proportion of the obsessed who are classed as insane was given as one third.

HEALING PRACTISED IN ALL AGES CONDITIONS OF HEALING— FITNESS FOR THE WORK.

Healers have exercised this function in all ages. It is enough for us to know that cures have been effected, but in precisely what way, or by what intrinsic agency, although very desirable, is of less consequence than the fact itself. Undoubtedly the law which governed them in all former ages is the same which operates to-day; universal in its application, and eternal in duration. When Jesus was asked by what authority he did these things, he declined to answer the question propounded by the cavilling Jews. The increase in the number of those who are performing such cures in a greater or lesser degree has been remarkable: they may now be numbered by thousands. Public confidence in the efficacy of magnetism is sometimes retarded by lack of power in the person claiming the ability to practice it, and sometimes by reprehensible conduct, but it cannot on either account be gainsaid, nor its claims set aside. It is not expected of any one healer that he can be successful in every case, but always in the variety of cases to which his magnetism is adapted. Water cannot flow above its fountain without the application of force, neither can a healer exceed the source of his power. Even Jesus did not claim to possess all power. He claimed to be in *rapport* with divine and powerful intelligences. At one time he says—so the record tells us—that he can call twelve legions of angels from the Father to aid him. If that power was vested in himself, wherefore the need of calling twelve legions of angels?

Certain conditions are to be complied with in order to render the efforts of the healer successful. We must not expect fruit from a tree, of a kind which it is not capable of producing. So it is with healers. "By their fruits ye shall know them." The highest benefits are not to be derived through those of them who have not sufficient regard for their own individual condition. Those who are not in harmony with themselves cannot be expected to harmonize others, nor meet with the best success in removing disease. Conditions have much to do with the curative process. Where the patient has confidence in the efficiency of the treatment, it is better for the operator to use the magnetic manipulations in quiet, and alone. In some cases the presence of other parties complicates the effect; and the outside force, foreign to that of the magnetizer, disturbs the latter, and even, with sensitive persons, destroys the good effect of treatment.

It has been questioned whether the beneficial effect of the vital magnetic treatment is permanent, and the fact has occasionally been denied. We may reply that for the reason that it is natural, and therefore in harmony with physiological laws, it is more likely to be permanent than artificial methods. But experience settles the fact of enduring good results. It is not to be supposed that when health has been restored, the patient will thenceforth be exempt from the ordinary causes of disease. If, for example, he is imprudent after recovering from an attack of Catarrh, or is accidentally exposed to such atmospheric changes as cause it, he will, like other persons, be attacked again. It requires ordinary care on his part, in avoiding the known causes of disease. He cannot knowingly or carelessly subject himself to unhealthy influences, and still enjoy immunity from sickness. Attention to diet, the use of bathing at proper times, and the observance of all hygienic laws is necessary. He will then be better able to resist the encroachment of other diseases, than if he had been medicated with drugs. His physical body will be preserved from premature decay, and his spirit from suffering. Healers generally operate upon their patients as they are impressed or moved to do. It is seldom that the manner is twice alike. Sometimes gentle passes are made; at other times the movements are more powerful, depending, as the impression is given, upon the condition and requirements of the case undergoing treatment. They differ with different operators: each one grows, so to speak, into his work. So that no general rule can be laid down for all to be governed by, but each will be gradually fitted to practise his own particular method.

Prudence is necessary to be exercised before the practice is adopted as a regular occupation. The gift should be sufficiently tested before its utility can be relied upon. He who has discovered by unmistakable signs that he possesses it, should be satisfied that he can do more good by exercising it, than in any other way. He must satisfy the public also that the power is sufficient to justify him in going forth in a public capacity. Many are now putting forth their pretensions, who do not possess the power in a degree sufficient to bring them the means of support, because their usefulness in treating the sick is too limited. When they are thus unable to sustain themselves, their failures injure the cause in which they are engaged. Others again possess sufficient power, but it is wasted by neglecting to bring it into use. They thus not only fail to discharge the trust committed to them, but injure their health by a perversion of the vital forces, which need their appropriate exercise in restoring the sick and suffering to health. Besides those who thus knowingly neglect to exercise their functions, there are again those who possess the power to a large extent, but lack the natural ability to put it into practical use, and it is in like manner wasted.

Persons who are endowed with the gift in a limited degree, as well as some of those who are more richly favored, can use it in a private way among their own friends, without the risk of abandoning their ordinary pursuits, and depending upon it as a business. When it has been proved that it has been sufficiently developed to yield them the means of support, and to do a positive good to others, they may safely venture to announce themselves as ready to serve the public. These remarks are made in a fraternal spirit, for the benefit of those who are in doubt, and have allowed their talent to lie dormant for years, being too timid to venture upon its trial.

The gift is often withdrawn, or remains dormant for a time, and is again made manifest with renewed and increased vigor. The fruits of study, and the experience of others is beneficial in preventing errors, particularly on the part of those who imagine themselves called to a "mighty mission," in regard to healing the sick; for such have too often failed to be of any benefit to others, or to accomplish any thing for themselves. Many have neglected, and at last abandoned a well established business, and engaged prematurely in the new calling, the prospect of which was precarious and uncertain; exhausted their surplus funds in travelling and other unavoidable expenses, and at length, instead of establishing themselves in business, have been compelled to return to their former occupation, with increased obstacles to surmount.

Of the many who believe themselves prepared for this new sphere of usefulness, it may be truly said that "but few are chosen." The reason is either that their development is not complete, or they do not live a true life. The work cannot be done with profit to themselves, or benefit to humanity under such unfavorable conditions.

What is most needed now is that physicians who are engaged in the practice of medicine should fairly investigate the method of curing disease by psychological power, or vital magnetism: and when they have ascertained that they themselves possess the latent power in any degree, to be courageous enough to cultivate it, and not be ashamed to use it, when by so doing, they can benefit their fellow-beings.

It is here proper to designate those who may employ their magnetism for medical purposes, as distinguished [illegible] others who should refrain from it. When it is [illegible] that a person magnetizing imparts more force [illegible] is received to supply the waste, he should desist from the pursuit of it as a regular occupation, but may with propriety continue to employ it in occasional instances.

Healers should not be actuated by a mercenary motive. If they have no higher incentive than the receipt of the dollar for treatment, it is evident that they have no proper calling for the work. They should always be prompted by a sense of duty, and derive pleasure from the good which they can do. They should not undertake the management of a case when they know there is no prospect of relief, nor continue treatment when there is no benefit derived, although they may at the beginning have had reason to hope for improvement. As much judgment is needed in applying magnetism as in the administration of medicine.

If physicians would employ magnetizers in cases which cannot be benefited by ordinary medication, they would exhibit an honorable intention, and increase their usefulness in a material degree. Some are already doing so, and are reaping a reward in an increase of their business. Patients have confidence in such men. When healers work in harmony with progressive physicians, it is better for both. We have often seen cases which would not yield to medicines, begin to improve after a single application of magnetism, and continue to improve rapidly, as if the body had been invigorated with new life.

There is frequently too much jealousy exhibited among healers, a propensity which ought not to be indulged. No one's magnetism is adapted to all cases and therefore no one alone can accomplish all that the vital power is capable of. As they differ in kind and degree, one can succeed where another would fail, although equally successful in his own way. In a variety of cases a variety of development in the use of magnetic power is to be distributed among them; each magnetizer finding his adaptation to one or more of them. Let each do what he can, leaving out self as the first consideration, in all cases avoiding inharmony: and if antagonistic influences become manifest, he should remain quiet until prepared to overcome them.

If the magnetism is not adapted to the case, it is likely to do more injury than good. The patient, if left to himself, can soon ascertain whether he is being benefited in such a degree as indicates an adaptation of the magnetic life force. If there is repulsion, or want of confidence on the part of the patient, it is of but little use to attempt to continue the treatment; as the beneficial results would be so slight as to be scarcely perceptible, if not absolutely useless, or on the other hand, injurious.

In some cases the healer is obliged to remain passive; at other times he must be active or positive: the influence which produces these conditions is exerted at the time of treatment.

We have already alluded to the practice of deception and imposture by persons professing medium powers. Some of these appropriate the name of magnetism, in order to introduce themselves to public notice; the treatment which they then adopt being as foreign to it as darkness is opposite to light. Such conduct should be exposed, and the guilty parties should be refused recognition among healers. Their motives are no more entitled to a charitable construction than if they were practising any other deception upon the community. It should be considered a moral duty to make the practice of magnetism respectable, and its honest devotees should seek to rid the country of pretenders who sail under false colors for the sake of gain, by showing the falsity of their pretensions.

There is a class of pretentious speculators who by adroit management contrive to get an advertisement of their pretensions inserted in the religious newspapers, while the publishers refuse to allow a word to appear concerning magnetic treatment. This class consists of Reverends, or persons who profess to have a very valuable recipe for some special disease; and desiring to aid the cause of humanity, will send it free of expense to any one who applies for it. When application is made, the formula is sent, accompanied by a circular announcing that to accommodate correspondents, they will send the medicine prepared for use, on the receipt of ten dollars. On going to a druggist, it is found that the prescription is so mystified that it cannot be compounded as directed, or that some of the ingredients cannot be had; therefore if they wish to get the benefit of it, they must send the money for the manufactured article. Thus the parties really benefited are those who have hypocritically pretended to be desirous of benefiting others, and the publishers who advertise the swindling scheme. Such tricks deserve exposure, instead of being aided and abetted by those claiming to be so wise and good that they must refuse to insert the business card of a magnetizer, on the ground that they did not understand it, and that it might lead their readers astray, or afford them an opportunity to investigate for themselves. What inconsistency!

Akin to these are the itinerant doctors, who travel over the country giving free lectures, and free medical advice. To the uninitiated it is marvellous how they manage to pay their bills; but if they will present one of the prescriptions to be compounded at a druggist's, they will soon learn that the price paid will compensate both druggist and doctor, and still leave a large profit. Such deception is at once apparent to a practical business man; but the unsuspecting can easily be duped, and the pretender escape` detection.

Returning to the qualifications of healers we may further say that when they can work together in harmony, a stronger concentration of force will be brought to bear on the patient; a result which is as reasonable to expect, as in relation to the combination of physical forces, or concentration of any one kind of power. But while a number of forces of a congenial kind are more energetic than a single one, it is the reverse when they are in antagonism. Hence when these magnetic forces do not operate harmoniously in combination, but being discordant influences to the patient, it is much better to depend upon one alone. The object to be gained in the healing process is harmony: various means are resorted to, for the purpose of harmonizing; some resort to the utterance of prayer, others require music to be played, or seek the tranquilizing effect of singing. Others remain passive, depending upon the power itself which operates through them, to harmonize the patient. Anything is beneficial which tends to restore the nervous system to a state of equilibrium.

It is injudicious to ask the patient, after treatment the question "Are you cured" or "have you been benefited?" inasmuch as it directs his thoughts to his morbid condition, and he is more liable, in imagination, to exaggerate its severity. It is better on the other hand to divert his attention from it as much as possible.

The patient should be imbued with confidence in the efficacy of the means that are being employed for his benefit, and the result will follow in its own good time. As soon as a change for the better is felt, it will be manifest to the magnetizer, from its vivifying effect externally; and it will be spontaneously announced by the patient.

Patients have sometimes been injuriously affected from fear, their medical advisors telling them that they are suffering from some serious ailment, which often turns out to be incorrect; that they are in a very low condition, and so on. It is better to encourage than to discourage in any case, for "while there is life there is hope."

It is improper to speak disparagingly of any treatment that is being pursued, unless it is a downright imposture; inasmuch as whatever of good there may be in it, is in some measure counterbalanced by discouraging words. If the patient, who is most interested, has confidence enough in any mode of treatment to try its virtue, provided the experiment affords any reasonable hope of complete or even partial success, let it be continued as long as he feels that it is doing good. When his judgment is enfeebled however, so as to be unreliable, and the friends have definitely ascertained that it is inefficient, they should advise a change. The patient should always be kept as positive, hopeful and cheerful as possible, which will assist nature to regain a healthful condition. Sleep is a great restorer, and should be induced by natural means. When caused by narcotic substances, the result produced is not so tranquilizing followed by invigoration, as that which follows the soothing operation of a natural process. It has been said that two hours' rest when awake is not worth much more than one hour of sleep.

The qualifications of nurses should not be passed over, especially when patients rendered irritable from a disordered state of the nervous system are placed in their charge. Sick persons who are naturally sensitive in health, are acutely so in a state of disease; it often in such cases amounts to morbid irritability. In such a condition they are prone to say and do things that they would refrain from when in the enjoyment of health. They are not then responsible for irritating words or unreasonable conduct, and therefore require from practitioner and nurses a large share of patience and charity.

Persons naturally irritable, who have not good command of their temper, are not suitable to be intrusted with the charge of the sick. Ill temper exhibited in passionate language and conduct is injurious to the health of the nervous system, and is often the means of producing sickness. Hence it is for the good of all that a calm and collected manner should be preserved, even in our intercourse with those in health, and more especially with those that are sick. We have known instances of anger agitating the nervous system so seriously as to result in disease, which nothing but the power of magnetism could allay.

The selection of persons for the duties of taking care of the sick is an important matter. The nurse should be congenial with the invalid. Indeed, the sick should not be in charge of either physician or nurse, who is not in harmony with them, who does not enjoy their confidence, and in whom there is no magnetic adaptation. Calls from friends during sickness, who are similarly antagonistic, should also be avoided, unless a request to receive the call should be made by the patient. Visitors should inquire, before entering the sick room, whether it would be agreeable for the patient to receive them. How often have we heard patients say of visitors after a formal call, that they thought kindly of them, but they never felt so well after their visits; yet they dare not say so to their friends, lest they should give offence. Others, in no wise superior to the inharmonious ones in moral worth, but whose life forces are different, may sit and converse with the patient, and after they have gone, the latter feels that new strength has been imparted, and a soothing effect produced upon the nervous system. Persons antagonistically constituted affect the sick unfavorably by merely entering the room, and the effect continues for hours, and perhaps for days. On the other hand, an incidental call from one whose forces are in a state of adaptation, will impart a beneficial influence, which perhaps may be the means of recovery. Physicians should acquaint themselves with the laws which govern these strange facts, and explain them to their patients.

The same persons visiting another patient might find the state of things reversed, the one whose influence was found beneficial before, now retarding the recovery, and vice versa. Again, they would respectively affect patients differently at different times. Both are presumed to be equally solicitous for the welfare of the patient, but their magnetic constitution or temporary magnetic condition differs from each other. Their good qualities in a moral point of view are equal: hence no blame is to be attached to the one possessing an antagonistic or injurious magnetic force.

HEALERS' PECULIAR MODES OF TREATMENT—POSITIVE AND NEGATIVE MAGNETISM—QUANTITY OF MAGNETISM REQUIRED.

For reasons already given, it has been stated that no uniform rules can be laid down for the guidance of magnetizers, as to the frequency or the manner of magnetic manipulations. Each has his own particular method. We may state the various manipulations and movements commonly used in our own, while operating upon patients, but these could not be adopted by any one else, and be implicitly followed, with the expectation of realizing the same results. Generally, to give an impetus to the vital force, we operate upon the base of the brain, and the spinal column. Then we place one hand on the pit of the stomach, and the other on the back, directing the patient to place one foot on one of ours. The position of the hands as to right and left is immaterial; either one may be placed on the back; though some magnetizers have a choice in this respect. The advantage of this, if any, can easily be tested. It has been supposed that the right hand was positive, and the left negative. As far as we can judge from experience, it makes no difference; and this has been confirmed by good authority. Either right or left can be made positive at the will of the operator. After sitting for a short time, we make passes from head to foot, for the purpose of equalizing the forces. When the system is torpid, we place one hand upon the head, and with the other make slight concussion on the back, gradually increasing the vigor of the strokes to the extent that the patient can bear. We make repeated concussions also on the palms of the hands and the soles of the feet. Such stimulating manipulations as the last named have a tendency to relieve the head by drawing the circulation downward.

We also make muscular compression upon the limbs by grasping the muscles of the larger limbs and the entire circumference of the smaller ones, which is done in rapidly repeated movements along the entire length in certain cases.

Where it is practicable the operations should be made by direct contact with the integument, but in many cases delicacy requires that a thin garment should be interposed. Silk should be avoided, for the reason that it is a non-conductor, and its use would tend to diminish the magnetic effect.

Others operate quite differently, and could not succeed so well with this method. It is generally given to the magnetizer what he shall do and say, at the time of using the manipulations; therefore he should yield to the impression, and operate as he is moved to do. Many use a tube, breathing through it energetically, which seems to invigorate the patient. This is an ancient movement; "and he breathed upon them." The power of the breath upon the sick, thus imparted, is sometimes wonderful.

Oftentimes all that is required is to sit quietly, holding the hands of the patient, and soon there is a pleasant glow immediately felt from the vitality imparted.

Among the variety of movements employed for the cure of disease is the kneading process, which can be practised by the patient himself or by a friend, but it is made more effectual when performed by a person of strong magnetic power.

What is called the swedish movement cure, the lifting cure, pinching, walking, and similar mechanical processes, all of which excite the various organs to increased activity, and accelerate the circulation of the blood, are highly beneficial to persons who are engaged in sedentary occupations. Where the mental powers are over-taxed, and the bodily organs have become torpid for want of suitable exercise, physical manipulations performed by a person endowed with magnetic power, as well as muscular movements made by the patient, are indispensable to the preservation of health, and its restoration when partially destroyed. Harmony between the mental and physical is one essential condition of health, hence good care should be taken to preserve it. A state of depression is inharmonious, as well as that of undue excitement. Moderation in all our conduct is necessary to avoid both extremes, and to secure the enjoyment of good physical health.

The sedentery are much troubled with a costive habit, arising both from want of good air and exercise, and from errors in diet; a condition which cannot continue for any length of time, without materially disturbing the general health. Proper food should be enjoined, and the kneading process may be resorted to in the commencement, which will prevent temporary disorder from becoming confirmed disease.

The ordinary battery is in some cases quite useful, but as its force is generated by the use of chemicals, it is more crude in its operation than human magnetism, and vital electric force; but it is found to be more effectual when employed by a person who possesses the other also, than by one deficient in that power, showing that the effect is modified, or increased, and that much of the good that is ascribed to chemical batteries really proceeds from human magnetism.

Those who employ such means, attribute whatever good results may attend their treatment, to the mechanical or chemical operations exclusively. The mechanical force they use through others, selecting any person of good muscular development, who has considerable power of endurance. They do not recognize, and often do not know of any natural gift in modern times, or do not understand it; and meet with but partial success in consequence. If they would inform their minds in regard to the subject, select their operators with good judgment, from adaptation; persons who would not exercise the invalid beyond his strength; their success would be very much increased.

This is true also of Water cure establishments. All indeed, who are engaged in treating disease, will sooner or later acknowledge the potency of magnetic treatment, and adopt first principles. Then will their usefulness be seen and known.

In these remarks we make no discrimination in favor of any particular mode of manipulation, but aim to be impartial, relying upon the correctness of the general principles stated. The power in each individual magnetizer generally increases with use; the exhaustion of vitality is compensated for by a new influx proportioned to the need, hence it is sometimes surprising to observers who witness magnetic treatment for the first time, to notice the great draft on the vital forces which can be tolerated, without causing entire exhaustion, the recuperation being furnished from an inexhaustible, yet invisible source. New developments of the subtle element are frequently taking place.

We now proceed to notice the magnetic conditions known as positive and negative. From the yielding of morbid symptoms to their desire to do good, magnetizers are liable, in the early part of their experience, to suppose that the effect is produced by the exercise of their own positive will power; but where this idea has been entertained they have subsequently found that in certain cases they could produce a better effect by remaining passive as regards the exercise of the will, while making the magnetic manipulations. In such cases the cure is performed when least expected, and with less physical exertion, the power being of a highly spiritual nature, with but little activity of the operator's individuality to complicate it.

We have stated that certain magnetizers are more successful than others in treating some forms of disease, showing that the potent elements which they derive, or which are attracted to them for the purpose, are precisely adapted to the requirements of the case, and that the operation of the force accords with natural law. If it were otherwise, all could do the same things in the same way, in treating different forms of disease with different patients. There need then be no selection of magnetizers with a view to the adaptation of the force which is employed through them, as all would be equal in degree and quality of force, and therefore equally successful.

Whether disease is caused by over-indulgence of passion, or over-exertion of the mental or physical powers in any way, and if any particular organ has become diseased, the secret of cure is to reach the cause, and by its removal restore the diseased organ to a state of soundness, by the adaptation of the proper force, which is accomplished by its being imparted through the magnetizer, while he is unconscious of its operation. Frequently it is not known at the time of the treatment that any good has been accomplished, but soon after the patient feels that his bodily condition has passed through a favorable change, which change is of a chemical character, and that he has been vitalized—that new life and vigor have been imparted; and he realizes that the diseased organ has been restored to its normal condition.

There must be a positive and a negative. At times the patient is negative, at other times positive: when in the latter condition, the magnetizer must be correspondingly negative. When the parties are in these conditions respectively, a cure can be much better effected than if both were in positive condition. When persons are fitted for the function of healing, and are well unfolded for the work, they will be influenced, or impressed as to what is required, and will become positive or negative, as may be necessary in each particular case.

As soon as a patient can safely dispense with treatment, it should be gradually diminished, the intervals between the times of its application extended, and as early as practicable discontinued. It is better that he should be independent as soon as his condition is sufficiently receptive, in order to drink directly from nature's fountain.

The immediate effect of magnetism is stimulating: if applied too often, or too long at a time, the nervous system becomes accustomed to its impression, and it ceases after a time to produce as strong an influence as at first. The decline in its beneficial effect would be the same as that of any stimulant. Its discontinuance must be gradual, to avoid the depressing effect which is observed on the sudden withdrawal of ordinary stimulants, to the use of which the system has become accustomed.

The quantity of magnetism to be employed is a matter of practical importance. In many cases it is necessary to avoid its too frequent use, in order that it may have proper time to diffuse its effects throughout the physical system of the patient, on the principle that a little leaven will leaven the bread, while a large quantity will spoil it. It has been demonstrated that those who have a surplus of either magnetism or electric force can aid others who are deficient in the one or the other, and at the same time benefit themselves. By the exchange of forces there is a mutual benefit. This is sometimes accomplished by taking hold of hands. In chemical magnetic force, the law of attraction requires that opposite conditions should be brought together, to secure the greatest benefit. The magnetic and electric forces regulate themselves, or are brought to a state of equilibrium on the same principle that the water of two adjoining ponds connected by a channel between them, will find its level. If the person whose magnetism is in excess happens to be sitting in the same room with another who is deficient in this respect, the change in their condition respectively will sometimes be brought about by their engaging in conversation, or in more marked instances by simply remaining passive, and perhaps unconsciously to both of them. Soon, however, the negative person, if unwell or fatigued, will feel revived. If unacquainted with the law of equalization, and by consequence unconscious of the change they are undergoing through its operation, they cannot realize it, nor will they give credit to the real cause of the benefit derived.

It is well known that public speakers, when their audience, or a portion of it is in sympathy or magnetically *en rapport* with them, can receive strength and mental sustenance, so that they speak with more energy and with greater effect. This fact is due to a law which as it becomes known, will be better understood, and its philosophical interest appreciated.

It is of more importance to be able to prevent the development of disease than to arrest it after incubaiton; and still more practicable than to eradicate it after it has by the lapse of time, run into the chronic stage. Often a single effort in the way of magnetic treatment will prevent a long, protracted severe illness. Oftentimes a change in the physical and spiritual surroundings of the patient, or a change of air or climate is of vital importance, both as hygienic and curative measures. So also with regard to bodily exercise, unless interdicted by debility. "A stitch in time saves nine," and "an ounce of prevention is worth a pound of cure," are old proverbs, which illustrate the value of preventive measures.

In magnetic treatment the force most adapted is that which is opposite to the force which characterizes the temperament and condition of the patient. As there is a natural law governing all life, it is useless for us to contend against it; on the contrary, it is much more satisfactory to work with nature, and in obedience to her laws. The Book of Nature should be studied earnestly, with an inquiring mind, for there it is that the student receives original ideas and thoughts, rather than from printed books.

The effects of magnetism upon different persons are not always alike. There is, indeed, as wide a difference as there is in the organization of the persons upon whom it is employed, and in the forms of disease with which they are severally afflicted. Sometimes it has had an effect similar to that of powerful medicine, the patient feeling unpleasantly affected by it for hours after the magnetization. The sickness seemed to be aggravated for the time being, but when reaction had taken place, it was perceived that there was a decided change for the better. At other times, when there is simply a reduction in the vital forces, the patient sinking down into a depressed condition, and there is no bilious derangement, the effect is to stimulate, followed by an obvious improvement from the first application of the power; with no injurious effect subsequently taking place, as is the case after the administration of stimulating medicines. It frequently produces the same sensation as hot water; again its quality is electric: cold chills run down the spinal column. Both are essential at times, that which is most appropriate depending on the patient whom it is designed to benefit. The magnetizer need take no thought about the quality of the power to be applied. It will be made known by the results produced. This seems to us to be the proper explanation of the variation in quality, though it may not satisfy all minds.

The time occupied in giving magnetic treatment varies with different magnetizers, as well as among the variety of patients. Some patients require more than others. Many magnetizers continue the treatment at each visit much longer than there is need of, fearing that if they were to abridge their manipulations, the patient would consider himself neglected; while in fact it would be better for both parties to limit the time to the actual need, which varies from one minute to more than an hour.

Care should be taken in regard to excessive magnetization in debilitated persons. But the length of time devoted to each visit may be diminished, and the employment of this valuable agency repeated at more frequent intervals, say as often as twice a day, or still more frequently if the result justifies it. In such cases also the operator should be cautious to avoid roughness in his manipulations.

We have alluded to the possession of the magnetic power in the domestic circle, where it could be used to advantage on the occurrence of sickness. There is probably an average of one such person to each family.

Certain rules are to be observed after treatment. With reference to the propriety of exercise, and exposure to the external atmosphere, after undergoing magnetic treatment, it depends on the nature of the disease for which treatment is sought, and upon the season of the year. Some patients know by intuition what is best for them. Others who have no intuitional promptings, nor knowledge of the subject, require advice. If the patient is in a sluggish condition, exercise should be kept up for a time after treatment. If he is naturally active and energetic, liable to exert himself to the point of exhaustion, he should remain quiet, so as not to lose the healing influence imparted. Generally a warm glow has been sent through the body, followed by perspiration. This renders the invalid more sensitive to cold than ordinarily, hence caution is necessary against exposure to a cold, damp atmosphere. In cold weather the patient should undergo treatment in his own room, especially if he be of a nervous, sensitive temperament, and negative in character. It is a bad practice for him, after receiving the glow of heat, and being stimulated to perspiration, to take a carriage, or ride in a horse car immediately, until he becomes chilled through. The benefit gained, is by such injudiciousness destroyed, and there is danger of absolute injury being done. We have seen the injurious effects of indiscretion on the part of patients, after perceptible good had been produced. Both healer and patient therefore, should exercise judgment in these respects.

DISEASES REMEDIABLE BY MAGNETISM.

In some forms of disease the magnetic treatment operates like a charm. First in order, we will mention *Inflammation*, which is so difficult to subdue by the use of medicines. Some magnetizers occasionally immerse their hands in cold water, before making their manipulations; and they are of opinion that they find it beneficial. With others there is no need of the water. *Rheumatism* in most cases is caused by the low state of the blood, or the want of vital force, which has usually been partially destroyed by a change of climate, or sudden fluctuations of heat and cold; injurious habits, such as high living, the use of intoxicating liquors, or over-eating of rich food and stimulating condiments. When the blood receives new life and vigor, and is chemically changed, it clears and purifies itself; and when this is done, the system is restored to its usual vitality. All curable disease of the blood can be subduced by a change of the qualities of the vital fluid accomplished through this power. All chronic diseases which are curable, with the exception of those requiring surgical aid, can be eradicated by simply imparting new life to the physical system. In some cases it is done suddenly; in others the bodily health is gradually brought up to the natural standard, as the flowers expand after a summer shower. *Fevers* have given way before the electrical power, sometimes in a single hour, in cases which have progressed for several days, without the least beneficial effect from ordinary medical treatment, showing a high degree of success.

In a lethargic condition of body, instances have been known where a sudden shock or fright has imparted a powerful impulse, stimulating to new life and action. For example a person has been confined to bed for many years. The house in which he has lived takes fire: in the excitement of the moment he jumps from the bed, and is well ever afterwards. Many magnetizers do the same thing by the exercise of their strong psychological will power. No doubt such cases could be cured by other magnetizers without the shock, by vitalizing the blood, thereby gradually changing the life forces. There are many persons whom shocks will not affect, but who by the latter means can be restored. There is nothing like magnetism for producing an equalization of the forces, and imparting energy; as for example when the hands and feet are cold, they are often restored to their natural warmth and glow by a single application of the adapted magnetism. Again, where a patient has not known for years what it was to perspire in his own person, after such an application the pores of the skin were opened, and the perspiration became as free and natural as in perfect health. The natural temperature of the surface of the body could then be kept up by ordinary exertion, and the system was in a condition to throw off diseased particles through the pores. It has been properly said that "if the pores of the skin were absolutely closed as with a coat of varnish, a person would not live ten hours. It would destroy life to keep in the system the refuse matter, which to the amount of a pint a day is discharged through these pores." This shows the importance of bathing, of sufficient exercise to keep the glands in good working order, and of frequent changes of the clothing that has to absorb this pint of matter a day. When the perspiration is checked or obstructed, either magnetic treatment, or a gentle sweat produced by simple means, will give an impetus to the life forces, and assist nature to regain her wonted condition. These simple but effectual means will prevent many severe cases of sickness if employed in season. *Consumption* is a form of disease in which great benefit is derived from this practice; it assists nature's efforts, giving strength to throw off effete matter, and recuperate the general tone of the system. Persons suffering from a debilitated condition of the lungs can also strengthen themselves by making slight concussions over the chest, increasing their force as much as they feel able to bear. When too feeble, this may be done by a friend under the proper conditions. The *Liver*, when laboring under any of its morbid conditions, is by this process aroused to healthy action, and given an opportunity to purify itself. The *Kidneys* receive benefit in the same way, and disordered conditions of the *Heart* can be relieved at once, and in a short time cured. Often the heart's functions are disordered only by sympathy with diseased conditions of other organs, and the free circulation of the blood is interrupted. This is regulated by magnetic force, after the application has been a few times repeated. *Dyspepsia*, and habitual *Constipation* are relieved in the same way. *Tumors* have been known to dissolve and pass away after magnetic treatment.

Magnetic treatment is highly beneficial in the diseases peculiar to females, and in the conditions which predispose them to disease. The pale and emaciated appearance of young girls accompanying derangement of the menstrual function, readily yields to its vivifying influence.

Many women become sick and prematurely old by over-taxing the function of maternity. Where this is the result of ignorance of physiological laws, and does not arise from wilful gratification of a sensual appetite, regardless of the consequences, a liberal dissemination of the necessary knowledge will abate the evil, and magnetic treatment will restore the strength and general health. The reproductive function should be under the control of the reason. No person should be the slave of another; but the woman has the natural right to the control of her person. A sense of moral equity awards it; and where reason and justice prevail, she will not be subjected to the abuse of a natural function.

The loss of health incident to the change of life which occurs in women is more readily prevented, and its restoration brought about more promptly by this method than by any other. The efforts of nature are assisted in carrying the patient through the critical period with safety and comfort. It is highly commendable in females to acquaint themselves with the laws of their being, by which they are enabled to avoid disease; and still farther to qualify themselves in such numbers as are necessary to treat the diseases of their own sex. Public sentiment is happily becoming liberalized in regard to the admission of women to the rank of practitoners, especially in cases of delicacy where the necessary examinations involve exposure of person.

Among diseases classed by physicians as tuberculous we may mention *Scrofula* as one to which historical interest is attached, in consequence of the long prevalent belief that cures were effected by the process of laying on of hands, provided it was done by monarchs, or under the authority of the church. We learn that the Pope conferred on Edward the Confessor, who was the last of the Saxon kings, (on account of his sanctity and devotion to the Christian religion,) the power of curing it in this way; from which it became known under the name of King's Evil. William the Conqueror succeeded Edward, and claimed the same power, but on what ground we are not informed. From that time until the reign of William and Mary, the gift was claimed by all the kings of England. Some of them derived enormous revenues from it at a shilling a touch. It is said that thousands availed themselves of the supposed benefit of this kingly prerogative. It was implicitly believed by men of judgment and education that the gift was possessed by the reigning monarch, and that he himself was free from disease, so that contact with him was esteemed a high privelege.

Dr. Willis, in a work published in 1687, remarks that "it is said the seventh son, or he that is born the seventh one after another in a continued series, can cure this disease by stroking it with his hand, and truly I have known many whom no medicine could help, to have been cured in a short time by this remedy.

"Few doubt but that this disease is wont to be cured by the touch of our king. The reason of such an effect ought to be assigned, not to any other thing than that in the sick the phantasia and strong faith of the hoped for cure induces strengthening of the brain."

In the reign of Charles the Second there were days set apart for the purpose of public healing, when all those who were brought before the king, to be cured of this affliction, were touched by him. Dr. Wiseman, who in the quaint phrase of those times, was styled the "Sargent Chirurgeon" to that monarch, says that he was an eye-witness to thousands of cures, and that it was well known, and could not be controverted, that the royal touch cured more in one year than all the surgeons in England in an age; and that strumous swelling of the neck, which had withstood the most powerful scattering and emollient remedies for many months, nay many years, had immediately disappeared after the royal touch.

This Court surgeon, who was an obsequious monarchist, maintaining the divine right of kings to rule and govern, gives most extravagant accounts of the wonderful deeds performed by his royal master, and speaks in contemptuous terms of those who doubt, though there were those who had the temerity to dispute and oppose the king's claims to the gift.

During the same reign a poor Irishman was strongly impressed with the belief that he had similar power. His efforts to cure scrofula by the laying on of hands were successful. His name was Greatrakes, and he became quite famous throughout his own country for his remarkable achievements. The fame of his doings soon extended to England: he was sent for by Lord Conway to cure his wife, which he effected by a few passes; after which several of the nobility were cured in the same manner. Greatrakes was of an honest and truthful character: imbued with the sacredness of his gift, and unselfish, he never accepted any reward.

There is an abnormal condition which it is proper to mention here, because of the remedy to be found in magnetism, viz: that disagreeable nocturnal state of unrest called "nightmare." Concerning this condition we have the following explanation. "It is a well known fact—to medcial men at least—that whatever tends to obstruct the free and natural flow of the fluids of the system, tends to produce a corresponding disturbance in the brain. And if the person afflicted be at all sensitive, or in other words, susceptible, at such times the spirit will partially retire from the animal life of the individual, and be able to take cognizance of the inharmony or disturbed pictures that are represented upon the brain. For, be it known, that all things—all circumstances in thought that pass the brain—are registered there, fixed there; and the spirit in its clairvoyant state is able to perceive these pictures. Sometimes they are exceedingly fair, sometimes the reverse.

Now it would seem in these cases, that there is some physical obstruction in the fluids, magnetic and electric, which is the cause of all these wild conditions. The very best remedy which we know of, may be found in magnetism, which restores, in a measure, what is lost; and removes the obstacles which exist in the circulatory system. Medical men will tell you that such cases are not rare."

Thus we might enumerate all the diseases that flesh is heir to, summing up with the general statement that the various forms of disease which are curable, are by this mode of treatment cured or benefited, after medicines have failed to produce any beneficial effect. In cases of *Paralysis* and *Insanity* it has not its equal for power to re-establish an equilibrium in the relative condition of the mental and physical functions.

While, as already stated, magnetizers should have confidence in the silent power or force which is employed through their several organizations, they should not bestow too earnest and anxious thought upon it, nor enter too much into sympathy with their patient, which has a tendency to depress their own vitality, without producing any increased benefit to the patient. The suggestions heretofore made, if carefully followed, will in due time enable the magnetizer to derive good results from his efforts.

The practice of commingling the magnetism flowing through two different operators, which is too often allowed in the treatment of a case, simultaneously or in immediate succession, is attended with injurious consequences. Sensitive persons should therefore avoid the conjoined use of two conflicting elements. They should be equally cautious in regard to promiscous circles, or assemblages of inharmonious persons. Such sensitives when undergoing treatment, have an intuitive perception of the forces adapted to their own needs, and can select such as will be beneficial, without the intervention of a third party. Without harmony in the forces, a magnetizer can do but little good. Insane persons, and those who are in a state of obsession are exceptions to this rule. Here the friends must intervene, and select such magnetizers as seem in their judgment, best suited to their condition; and who are in themselves harmonious.

The want of observance of the foregoing precaution has, in many cases, resulted disastrously. When a course of treatment has been in progress, and attended with good results, for some reason or other another magnetizer has been called in, who was not in harmony, or in a condition of adaptation; and the force peculiar to the latter has destroyed the good effect which had been produced through the agency of the former; and the final result was that no ultimate benefit was derived from either. It is not an infrequent occurrence that patients after receiving decided advantage from the application of this curative means, go out among other persons who are not in harmony with them, and the antagonistic influence which surrounds them, neutralizes the beneficial effect of the magnetic treatment, and the magnetizer is unjustly blamed for failing to effect a cure, or in any perceptible degree imparting benefit to the patient.

Another practical point to be observed is that feeble, delicate persons should refrain from over-exertion, resulting in fatigue, while undergoing magnetic treatment. Prudence is necessary in this respect at all times, but it is almost useless to attempt to magnetize under such a disadvantage, the result generally proving it to have been ineffectual.

As an aid to vitalized magnetism, some of the mineral springs perform a useful part. Many medicinal springs have latterly been discovered, which, perseveringly used, have been the means of relieving many cases of chronic disease. Besides those holding in solution a variety of mineral salts, others have within a few years been found to contain magnetic and electric properties in a high degree. One of these, discovered in the State of Michigan, is so highly charged that a knife-blade immersed in its water for a few minutes, becomes sufficiently magnetized to attract needles, and iron tacks, which can be held suspended as by an ordinary magnet. The magnetic property thus imparted is retained for months.

A discovery of such interest and importance, has attracted a good deal of attention. Many persons have flocked to the springs from all quarters, for the purpose of testing their medicinal qualities, and the result in a considerable number of cases has been very satisfactory.

Thus we see that in the great labaratory of Nature there are valuable forces already prepared for our use, which if properly applied, can be made a great blessing to the human family. They invite us to partake of the bounty spread before us as a gift from heaven, without money and without price.

With reference to the power of healing at a distance we quote the following from an intelligent patient whose veracity cannot be questioned, a lady of culture and liberal thought. "My health greatly improved after being magnetizd by you, and I must also inform you that whenever I felt unwell or fatigued, I would take your letter and hold in my hand; and often if I had pain in my head or chest, would place the letter upon the affected part, covering the same with my hand, and would feel relieved immediately. I felt the influence as readily as when your hands touched me, imparting that peculiar warm, soothing, magnetic glow and vitality."

NATURE THE SOURCE OF CURE.

The curative process is effected by the power of Nature, yet it is often the case that something else gets the credit. All that any plan of treatment can do is to assist her. That kind of practice which is not productive of direct injury, by creating other forms of disease, is the one most needed to supercede others which do not harmonize with nature. Magnetism assists nature when medicine fails. The cures performed by it have often been in cases which had been experimented upon by the use of nostrums, and by many of the well known remedies, until the disease had become chronic; and this was the last resort. Sometimes the disease is eradicated immediately; at other times the magnetic force gives an impulse, which sets the life current in motion, but time is required to effect the cure. In the latter, the physical system goes through a chemical change, the vital forces being increased more naturally than by any other known practice. Some of these cases have been pronounced incurable by some of the highest medical authorities. Such are not now uncommon. The plan of treatment is entitled to be ranked as a legitimate calling, certainly as much so as those which though more popular, are less effectual, and cannot compare with it.

In further illustration of the inherent material and spiritual qualities of man's being, and of the agency of the nervo-vital and spiritual forces in the cure of disease, we make the following appropriate quotations, which, although not occurring in consecutive order, will enable the reader to form a clearer conception of the bearings of the subject:—

"Man, as a physical being, is composed of absolute principles, the aggregate of which make up his spirituality. The spiritual principles are wholly dependent on the physical, and any attempt to ameliorate the condition of the race, must be founded on the amended condition of the bodily powers. Within the human organism resides a trinity of forces: the electric, galvanic, and mesmeric or magnetic; and the economy of Nature in the physical form gives also three currents: the arterial, venous, and electro-magnetic or nervous fluids—the uniform co-operation of each with each forming the basis of that state we call physical health, which is the true road to spiritual progress. The nervous fluid has its life from two organs in the form—the brain and the spleen; each of which supplies the food necessary for the proper support of the equilibrium of the nervous system. In years to come mortals will understand the hidden meaning of the law governing their being: then these powers will be put in command, and health will be the rule, not the exception—a health obtained by the cultivation or restraint (if need be) of certain naturàl principles, not by recourse to drugs, which poison the occult forces of the body.

"The law of physical harmony must be better understood by us; it is the duty of each one to investigate. Every medicine or article of food taken into the system which is not adapted to the case in hand, is productive of deleterious effects, and only by a knowledge of the requisite remedies can health be established, and we become true men and women. Can a man with a diseased stomach possess an equilibrium of temper? Can a woman with shattered health preserve a collected state of mind amid her manifold cares?

"Upon a correct state of the physical forces depend not only health, but also the power of receiving impressions. Thus it will be found, by a knowledge of the laws governing the nervous fluid, that a surplus of magnetism gives sustenance to the impressional faculties, a surplus of galvanism produces powers of healing, and a surplus of electro-magnetic or nervous fluid gives more power to decide upon those problems which meet us in life at every footstep.

"The spiritual nature is only mantled with the fleshly covering of the physical body, and yet it is so centered therein that it must have the proper quantity of spiritualized vitality it needs for its support, and this supply can only be obtained through efforts to cultivate a healthy physical organism, by the observance of physiological laws. As the time will come when physical suffering will yield to man's enlarged knowledge, so also will the spiritual nature be freed from the constant jars and shocks of to-day, and a calm, healthful serenity of body and mind will be the normal condition of earth's inhabitants."

"The body physical being possessed of two distinct sets of nerves, the voluntary and involuntary, science tells us that the action of the subtle nervous aura, or force, when passing over the involuntary system, causes involuntary action. In its play upon the nervous system, it acts upon the brain; its force is first applied there, and from thence it descends throughout all the voluntary nervous system. You may ask, "Is there any difference existing between the force that acts upon the voluntary and that which acts upon the involuntary nervous systems?" I should answer "No; I believe them to be one and the same power." We may give as many names as we please, but after all it is but one force. You call magnetism and electricity two distinct forces. This is a mistake—they are one. Seen under certain circumstances you call them magnetism—under others, electricity. These terms must be used by you so long as you have need of them; but as you go up in life, you will drop one after another, and come down to simplicity of expression. To-day you cannot understand it, so you must have your magnetism, your electricity, your psychological forces, and a thousand and one terms for one and the same thing.

"It is a well known spiritual fact that disembodied spirits are largely endowed with healing powers. Many of them have made this particular branch of science a special study, so that, under favorable conditions, they can perform what to your unenlightened senses would be called miracles—such as restoring those who are violently sick to immediate health—such as restoring those who have met with accidents—if, indeed, there are any such in Nature. All these interruptions to the harmonious course of Nature can be easily overcome by supplying that which the accident or disease has destroyed or removed. If the question arises, Has Nature any adequate supply?—to be sure she has. A large provision she makes for all these necessities; and it is only because you do not know how to make right application of them that you are sick—that you go about, day after day, complaining because of the heavy hand of disease that is laid upon you.

"Whatever condition of atmosphere will arrest the decay of the parts will perform the cure. It is of itself all the remedial agent that is necessary. Pulmonary consumption, we are informed, can only be cured by atmospheric means. No amount of drugs can by any possibility effect a cure. They may palliate for a time, but they cannot cure. Diseased lungs can only be restored to a state of health by atmospheric remedies. If Nature furnishes them in any locality, why Nature then stands there as a pre-eminent physician. The time is coming, I believe, but it is far distant from the present, when diseases of every kind will have departed from the earth, because the inhabitants will have learned how to prevent disease. Now they dwell in the midst of it, and court it, because they are ignorant concerning the laws of life; but as mind progresses it will pass out of this unhappy condition. There will come a time for the inhabitants of the earth, I do most firmly believe, when there will be no more disease, not of any class; when bodies will come into the world healthy and go out naturally. But, as I before remarked, that time is far distant from the present. But Nature is marching on to it perpetually. The earth does not make a single revolution upon its axis but it brings you nearer to that millennium.

"Having made yourself acquainted with yourself, go out into Nature, and become acquainted with that. Learn how Nature acts upon you, and how you act upon Nature. Then go a step higher, and learn what it is that animates Nature, and how that which animates Nature is allied to that which animates you. These are the studies that would be productive of the greatest amount of good to the human race.

"Nature has marked out her course for human life, and it is a very exact and wise course. As the body grows old or becomes burdened by years, it parts with its magnetic life, that it may the easier pass through death. Now suppose it retained all its magnetic vitality to the last moment of its earthly life; what would be the result? Why, the most terrible struggle between the magnetic and electric forces; consequently a very hard death. See how wise and humane Nature is to make the body part gradually with its magnetic forces, that it may pass easily through the change called death; and in your ignorance you ask to retain it. It would be the greatest of curses if you could."

CLAIRVOYANCE.

THIS faculty, or the gift of clear sight, (clear-seeing, as its derivation imports) depends upon conditions, as does the gift of healing. Therefore it is not infallible, any more than the latter; and in some cases, when the conditions are unfavorable, a trial of it will fail to give satisfaction. Some clairvoyants are apparently born with the faculty, or at least its exercise commences at such an early period in life, as to lead to the inference that it has existed from birth: in others it is gradually developed. Some claim that it is the spirits who take possession of their organism, who are clairvoyant. Whether the one or the other, both the classes of persons so favored are reliable at times; but they may sometimes meet with persons whose chemical life sphere is not always penetrable, and by all clairvoyants, therefore it should be used reasonably, and the facts as they exist should not be expected to be ascertained at all times with certainty. A great amount of good in detecting disease and its causes, where the diagnosis had previously been obscure, has been accomplished by means of it. We consider it as reliable as is human testimony in most cases: where the clairvoyant is fully unfolded, and the conditions are proper, it is far superior to any other source of knowledge.

The gift of prophecy is sometimes exercised by clairvoyants in a wonderful degree, some of the events foreseen transpiring with such exactness as to equal those recorded in the bible, the reliability of which is received by multitudes of persons without question. The laws of the Universe being fixed, eternal and unchangeable, why cannot the law which governs such phenomena operate just the same now as then. Those who seek for truth with an unprejudiced disposition, can find similar cases within the circle of their own acquaintance, which cannot be gainsaid nor invalidated.

Thousands rejoice in restored health in all parts of the country, for which they are indebted to the clairvoyant gift. The morbid condition is seen as clearly as if reflected in a mirror. At the same time a simple remedy is pointed out, the operation of which is in harmony with the natural efforts. Its effect being known, a change of remedies can be made when necessary, without experimenting in uncertainty.

The prevision exercised by clairvoyants frequently enables them to detect disease lurking in the system years before it manifests itself in definite form. It can therefore be made of great value in ascertaining the existence of suspected disease, when not fully developed, thus enabling the subject of it to adopt such measures as are necessary to arrest its progress.

A case of a lady came to our knowledge, wherein serious consequences were averted, if not life saved, by the timely detection of an important error, through her clairvoyant power. She had sent for an ounce of tincture of Senna: the Druggist by mistake put up tincture of Iron, and labelled it Senna. The color being similar, most persons, unless happening to taste the liquid, especially when having confidence in the Druggist, would have taken the medicine as directed. By means of the clairvoyant power the error was detected, and the consequences which would have followed the misuse of a medicinal preparation, good in its place, prevented. Extreme irritation, from what is stated of the properties of the medicine put up, would have been the result of an over-dose, and might, in the condition of the patient at the time, have brought about a fatal termination.

We have known prescriptions to be given when the prescriber was under the influence of this power, which were not what they should have been; and the sick person was, to say the least, in no wise benefited. Failures resulting in this way, from a too implicit confidence in the possession of a delegated power, should be a caution to clairvoyants. They should not assume it to be infallible, but when they have reason to believe that it has been in a good degree developed, proceed at once to test whether it is reliable in their own cases. Numerous truths have, at various times been given us through this singular power, and at times these have not been unmixed with errors. The latter we attribute either to unfavorable conditions surrounding the person claiming to possess clairvoyant power, or to a defect in the real development of the faculty in that particular person.

There are to be found amongst us in all walks of life, amid persons of all occupations, both the true and the false. It becomes us as intelligent beings, to exercise our reason, trying all things by the standard of truth, with the capacity of judgment that has been bestowed upon us, and proving the good that is in them. We should not condemn the wheat merely because of the chaff that is in it. We often find truth somewhat mixed with error. There must always be a genuine original, where there is a counterfeit. Clairvoyance is an established fact; so recognized by all candid investigators. So thoroughly are we convinced of its truth, that to us it seems a mystery that there can be found a thinking, intelligent being who doubts the fact.

Among the numerous cases which have occurred in all parts of the country, and indeed in nearly all the countries of the world, we content ourselves with the citation of the following, in illustration of the phenomena of clairvoyance. The first occurred in the city of Brooklyn, and through statements published in the daily newspapers of that city, during the last three years, acquired considerable notoriety. The testimony concerning the phenomena, coming from those who had never believed in the existence of such a faculty as clairvoyance, will be more convincing to the minds of doubting and skeptical investigators, than if the phenomena had occurred among, and had been testified to, only by believers, who had witnessed and recognized such phenomena before. When the connected and complete history of the case is fully made known, it will startle the community still more.

The subject through whom the manifestations have been made is a young lady, whose mental faculties had been over-taxed by hard and close study. She became debilitated, and lost her natural eye-sight. Then followed almost total loss of appetite. She could partake of food or drink only at long intervals; the juice of grapes being her only source of nourishment for three years. Such is the report of her attending physician. We learn that she is still in the same condition that she has been in for that long period, lying in bed with one hand under her head. While in this condition she executes crochet work of the finest quality in worsted, blending the colors as accurately as if in the exercise of her natural sight, and manipulating her work with as much dexterity as others in the enjoyment of ordinary health. She makes also the finest styles of wax flowers. At one time she wrote twenty verses of poetry in twenty minutes, the penmanship being equal to that of our best writing masters; and the sentiment of the poem of a high order. She will answer sealed letters without opening, as rapidly as they are given to her. She has described to her visitors their daily acts when absent, and indicated the exact time of day, which was proved by referring to a watch. These things are done while her eyes are closed to external objects. While under this influence, she has written a large quantity of manuscript, which we hope to see put in print in due time. Coming unsought, and among those who had never understood any thing of clairvoyance, it is in itself more convincing, though not more truthful, than a thousand other developments of this nature.

The case of Heinrich Zschokke, although frequently quoted in illustration of a remarkable psychological phenomenon, possesses unabated interest because of the excellent character and distinguished position of that author in the world of letters, and its being among the most marked of the earlier modern developments. It is here reproduced from his *Autobiography*.

"It has happened to me sometimes, on my first meeting with strangers, as I listened silently to their discourse, that their former life, with many trifling circumstances therewith connected, or frequently some particular scene in that life, has passed quite involuntarily, and, as it were, dream-like, yet perfectly distinct before me. During this time I usually feel so entirely absorbed in the contemplation of the stranger life, that at last I no longer see clearly the face of the unknown wherein I undesignedly read, nor distinctly hear the voices of the speakers, which before served in some measure as a commentary to the text of their features. For a long time I held such visions as delusions of the fancy, and the more so as they showed me even the dress and motions of the actors, rooms, furniture, and other accessories. By way of jest I once in a familiar family circle at Kirchberg related the secret history of a seamstress who had just left the room and the house. I had never seen her before in my life; people were astonished and laughed, but were not to be persuaded that I did not previously know the relations of which I spoke, for what I had uttered was the literal truth; I on my part was no less astonished that my dream-pictures were confirmed by the reality. I became more attentive to the subject, and when propriety admitted, I would relate to those whose life thus passed before me the subject of my vision, that I might thereby obtain confirmation or refutation of it. It was invariably ratified, not without consternation on my part. I myself had less confidence than any one in this mental jugglery. So often as I revealed my visionary gifts to any new person, I regularly expected to hear the answer, 'It was not so.' I felt a shudder when my auditors replied that it was true, or when their astonishment betrayed my accuracy before they spoke. Instead of many, I will mention one example, which pre-eminently astounded me. One fair day in the city of Waldshut, I entered an inn, (the Vine,) in company with two young student foresters; we were tired with rambling through the woods. We supped with a numerous society at the *table d'hote*, where the guests were making merry with the peculiarities and eccentricities of the Swiss, with Mesmer's Magnetism, Lavater's Physiognomy, &c. One of my companions, whose national pride was wounded by the mockery, begged me to make some reply, particularly to a handsome young man who sat opposite us, and who had allowed himself extraodinary license. This man's former life was at that moment presented to my mind. I turned to him and asked whether he would answer me candidly if I related to him some of the most secret passages of his life, I knowing as little of him as he did of me? That would be going a little further, I thought, than Lavater did with physiognomy. He promised, if I were correct in my information, to answer frankly. I then related what my vision had shown me, and the whole company were made acquainted with the private history of the young merchant; his school years, his youthful errors, lastly with a fault committed in reference to the strong box of his principal. I described to him the uninhabited room with whitened walls, where, to the right of the brown door, on a table, stood a black money box, &c. A dead silence prevailed during the whole narration, which I alone occasionally interrupted by inquiring whether I spoke the truth. The startled young man confirmed every particular, and even, what I had scarcely expected, the last mentioned. Touched by his candor, I shook hands with him over the table and said no more. He asked my name, which I gave him, and we remained talking together till past midnight. He is probably still living!

"I can well explain to myself how a person of lively imagination may form, as in a romance, a correct picture of the actions and passions of another person, of a certain character, under certain circumstances. But whence came those trifling accessories which no wise concerned me, and in relation to people for the most part indifferent to me, with whom I neither had, nor desired to have, any connection? Or, was the whole matter a constantly recurring accident? Or, had my auditor, perhaps, when I related the particulars of his former life, very different views to give of the whole, although in his first surprise, and misled by some resemblances, he had mistaken them for the same? And yet, impelled by this very doubt, I had several times given myself trouble to speak of the most insignificant things which my waking dream had revealed to me. I shall not say another word on this singular gift of vision, of which I cannot say it was ever of the slightest service; it manifested itself rarely, quite independent of my will, and several times in reference to persons whom I cared little to look through. Neither am I the only person in possession of this power. On an excursion I once made with two of my sons, I met with an old Tyrolese, who carried oranges and lemons about the country, in a house of public entertainment, in Lower Hanenstein, one of the passes of the Jura. He fixed his eyes on me for some time, then mingled in the conversation, and said that he knew me, although he knew me not, and went on to relate what I had done and striven to do in former times, to the consternation of the country people present, and the great admiration of my children, who were diverted to find another person gifted like their father. How the old lemon merchant came by his knowledge he could explain neither to me nor to himself; he seemed, nevertheless, to value himself somewhat upon his mysterious wisdom."

Vast progress has been made since the time when the occurrences just mentioned took place, with reference to the practical utility of clairvoyance. For the purpose of detecting disease and obtaining remedies, it has been extensively resorted to. We find that clairvoyant physicians who prescribe the use of medicine generally prepare it themselves. Without doubt it is by this means made to partake of their own magnetic life force, and is more beneficial than if prepared by the patient or friends. Simple, harmless medicines prepared by those in whom the patient has confidence, are more effectual than others, in themselves more powerful, prepared by persons with whom the patient has no sympathy, and whose magnetic forces are antagonistic to each other.

Medical practitioners are now frequently making it a point to consult clairvoyants when they have difficult cases, which resist the potency of ordinary medication. Some are timid, and try to keep it a secret; gradually however they become willing to acknowledge it, and at length not considering it a disgrace, are emboldened to defend the rectitude of their course. Would that there were more who dare to be governed by their own sense of right; even at the risk of rendering themselves unpopular. There is need of educated physicians as well as surgeons, who are liberal minded; the one with a scrutinizing perception, studious inclination, and critical judgment, the other with mechanical genius, aided and directed by study. There is a difference between physicians in regard to natural fitness, independent of education. One with the same education, yet no greater general scope of intellect, will prescribe more skilfully than another, and consequently with greater success; and the profound in learning, other things being equal, will be able to accomplish more than the superficial. So that at last, they are adapted for their work under the operation of a natural law which lies at the foundation. The superiority in the one case over the other, is to be accounted for, more by the difference in magnetic qualities than by any other circumstance.

The benefit of clairvoyance in surgery is illustrated in a case which had been pronounced cancer, the physicians giving it as their opinion that it must be removed by a surgical operation. The lady who was the subject of the affliction called upon a clairvoyant, whose advice was against its removal by the knife. A magnetizer was subsequently employed, whose manipulations dispelled the disease; and the physicians were candid enough to acknowledge the cure. Numerous cases of a similar character might be cited, wherein disease has been eradicated by the use of magnetism, and thus prevented a resort to harsh measures, which would have been inevitable, had the utility of clairvoyance and magnetism been denied, and the benefits of the latter been refused. In the case mentioned there was no cancer, but a tumor which resulted from a morbid condition often occurring in women at the change of life. As soon as relaxation of the tissues took place, the tumor disappeared through the operation of a natural process. Had it not been thus dispelled, it might have terminated in the full development of cancer.

Another case of mistaken opinion occurred under our observation. A man was suffering extreme pain from an obscure disease, for which he had been under the treatment of three or four prominent medical practitioners: no two of them agreed upon the cause. The case had been protracted several months. The patient's limbs and body were considerably swollen; he had spasms about every three hours. A magnetizer was called in: after employing his treatment a few times, the swelling of both body and limbs abated, and they were very soon reduced to their natural size. The spasms discontinued, and all that remained that was unnatural, was a tumefaction situated just below the heart. During the magnetic treatment, the general swelling did not return. One of the physicians expressed the opinion that there was cancer of the kidneys, and that nothing could save the patient. The magnetizer discontinued his passes, the physicians having in the mean time directed quieting medicines to be given, which did not operate harmoniously with the magnetism. In a few weeks the swelling again took place, the patient lost his vital force, and his spirit left the form. It appeared that a large solid fleshy tumor had pressed against the heart as the swelling increased. The post mortem examination showed that the tumor was the cause of suffering, but it revealed no cancer. Thus it is evident that physicians, notwithstanding the knowledge they possess, will sometimes err in opinion, and mistakes in practice will occur, as a necessary consequence. We do not mention the names of those who were concerned in the management of the case, as we wish to avoid personalities.

In another case which had been under the care of a physician for six months, medicine failed to reach the vital forces, and a magnetizer was called in. After repeating his treatment a few times, the condition of the patient was materially changed for the better—the surface from a deathly cold becoming suffused with a pleasant warm glow, soon followed by perspiration. The attendant thought that one of the powders which had been given by direction of the physician, when in a cold inactive state, might aid in inducing sleep. The circulation being so much more active and life-like after magnetic treatment than before, the medicine had more effect; and after taking the morphine powder, he slept to wake no more in the body. This shows that medicine is at times inoperative, as in this case, when the system is in a cold, inactive state; but after the stimulus of magnetic treatment it has its usual effect.

One other case is sufficient on this point. A young lady was suffering from excruciating neuralgic pains. Her family physician, and a professor in a medical college had been treating her for two weeks without any beneficial effect. Blistering and a free use of chloroform had been employed, which cost the family some twenty five dollars. A magnetizer was called, and relief was oblained immediately. The explanation is that new life and vigor was imparted to the patient, making a decided chemical change in the condition of the life forces. If these are facts, it is not surprising that physicians should express astonishment when they perceive the evidence of a latent force in nature, capable of producing powerful effects in the hands of persons who have never given the least attention to the study of medicine; and in cases in which they had exhausted all known remedial agents.

We would not desire to dictate to any one what particular method of practice he should employ when sick. He may take the contents of a drug shop if he chooses;

but we claim for the practice of magnetism equal rights and priveleges. Let each mode of practice stand upon its own merits, without restrictions or special favors. And he who changes his opinions from honest and intelligent conviction, should be allowed full liberty to give them expression in this land of the free. It has been well said that "He that confesses that he has changed his mind confesses that he is wiser to-day than yesterday."

We have heard of physicians in whom the magnetic power had been developed, and who so acknowledged to their patients, yet when one of these was suffering in the extreme, and their prescriptions failed to afford relief, refused to call into the service of their art this natural gift. The reason given was that it was not according to their school of practice; and they did not want to do any thing in the way of treating disease by this method, even if they knew it would give relief. In one instance, where the physician himself was sick, and the benefit of the magnetic treatment was offered, he refused it, not only on the ground already stated, but also because it conflicted with his religious belief; alleging that if he received benefit from its use, it would help the cause, and he would rather die than employ a professional healer. At the same time the male members of his church formed themselves into a band of healers or "rubbers," two of their number being selected nightly to watch with him. This was continued for many months, until "death" came to his relief. All the relief that he obtained from any source during his last illness was from the manipulations of these voluntary nurses, whose services he availed himself of, and whom he desired to rub him almost continually. In such a case powerful magnetic force, exerted through a despised healer, is competent to give relief, and often in a short time to effect a permanent cure. "Consistency, thou art a jewel!"

THE MATERIAL AND SPIRITUAL BODIES —THE CHANGE CALLED DEATH.

It was intended by the Author of our being that the material life of mankind, like all things in Nature, should endure until its offices have been fulfilled. So if the physical body were not prematurely destroyed by accident or disease, man would continue to inhabit his body until a ripe old age; and when the change called "death" takes place, it would be hailed as a welcome messenger. The house we now live in will be needed no longer. It will have served its purpose, as the dry husk enveloping the ear of corn, which has ripened for the harvest. The event will come to us as naturally, and appear as much in its proper order, in the one case as the other.

In the present age, disease frequently takes the brightest spirits to the other life in an undeveloped condition, because of the deterioration which has resulted from a violation of hygienic laws. They are born into this world with an unhealthy physical organization. The spirit's condition then is like fruit, that from various causes drops prematurely from the tree. We are convinced that earth's experiences are needed by all; and that it was not intended for any spirit to exchange spheres of life in infancy, but that it should arrive at maturity—become fully ripe, fulfilling its duties, and deriving pleasure from the exchange when the proper period should arrive.

It becomes a serious moral question how far those who assume parental relations are culpable for the interruption of the natural order, by bringing into existence unhealthy children. It is well known that the various qualities of mind and body are transmitted from one generation to another. Hence it is unquestionably a great evil for any one who is suffering from a constitutional transmissable disease to be instrumental in thus entailing it upon their offspring, who can be of no comfort to themselves or friends, but must live a life of suffering and sorrow, pain and misery, until the change prematurely comes, that frees them from the material body. The procreative function should be under the control of the intellect and moral sentiments. Parents have themselves to blame, in a great measure, for the sickness of children in such cases. We cannot plant thorns and gather grapes. Each seed will produce its kind, so that we may know what to expect in the harvest. Good, sound fruit cannot be obtained from decayed trees, nor ripe fruit out of season.

Concerning the nature of life, and the relation between the physical and spiritual bodies, we make the following extracts from the utterances of one of the higher intellgences:

"I believe all life to be essentially the same, wherever it finds expression. That which exists in the mineral is related to us; that which exists in the animal, in the vegetable, and that which exists in the mineral possesses a distinct individuality even there. It is individualized by its surroundings. The diamond becomes the diamond by its surroundings. The life of it is precisely the same as is my life and yours. The infinite Spirit of Wisdom hath so ordered it that life shall travel up through all these lower gradations till it reaches the highest in the heavenly spheres. It finds an intelligent expression only in the animal creation. It becomes more intelligent as it rises into the superior, and still more intelligent as it rises into the celestial."

"Since the things of the spirit are eternal, and the things of the body are temporal, surely those of the spirit are most valuable. Those treasures that belong to the soul are more valuable than those that belong to the body. Health of the body, so far as this world is concerned, is good—a great blessing; but so far as the other life and things that belong to the soul are concerned, it is of no value whatever."

The event called *death* must come to all. It is a part of the great system of life, which could not be completed without it. It is not annihilation, but a transfer of the individual being from one condition or sphere to another. Since time commenced, growth and decay has been going on from generation to generation.

All go in turn to the spirit life; good, bad and indifferent alike, in obedience to the requirements of a fixed, universal law. The cures said to have been performed by Jesus and his Apostles did not extend life indefinitely. The change came to their patients at last, as it does to the sick and the time-worn to-day: they having finished their allotted time on earth, or shortened it by infraction of vital law, long since passed into other spheres. The acts performed in those days, which excited wonder and veneration, were designated by terms quite different from those which would be employed in the present age in describing similar acts. Where the record, for example, reads that persons were cured after being dead, it should read, or be interpreted to mean *supposed dead*, or *apparently* dead. We cannot entertain a doubt of the authenticity of the biblical statements of cures, for we see the same things done through the agency of the magnetizers of to-day. Thousands of cases could be cited, quite as wonderful as those which occurred in the past; and the grateful recipients of the benefits of treatment would be willing to stand up and testify of the great good that has been done for them.

PSYCHOLOGICAL PHENOMENA.

UNDER this head may be included a variety of interesting facts, and the mental reflections to which the contemplation of them gives rise. The influence exerted by one mind over another, the formation of individual opinion, the changes of belief, the power of the Will over disease, the activity of faith as a mental operation in the removal of disease, the power of the imagination, and other topics might be considered. But this opens a vast field of thought, altogether too extensive for our purpose. We must confine ourselves to such as have a direct bearing upon the general subject of vital magnetic operations.

We are constantly surrounded by a variety of influences, which have to a greater or less extent, a controlling power over the feelings and conduct. They affect sensitive persons in a manner as subtle as the needle is affected by a magnetic current. Some of them termed psychological affect such persons injuriously, and if not counteracted, will produce disease. Fortunately they are not uncontrollable: through the magnetic power of a person of proper adaptation, they can be changed with the greatest ease.

In one or other of the numerous forms in which it is manifested, it operates through certain persons as its active agents; in some unconsciously, in others by the exercise of the will power. When in possession of persons who are governed by correct moral principle, it is not only harmless, but can be used beneficially. Misused by unprincipled persons, as in the perversion of any other power from its legitimate use, it may be made the means of doing great injury. Of the latter class, those who possess it in an intense degree, are dangerous in the community, and should be shunned as one would avoid a poisonous reptile.

The more prominent varieties of this power may be arranged under three heads, viz: 1. that which is of the lowest order, being manifested through the animal, or selfish propensities; 2. that which is manifested through the intellectual faculties; 3. that which is derived from invisible, disembodied spirits.

1. The influence for which the lower order of propensities of the human organization is chosen as the channel of development, can be made use of, as gunpowder or steam is; but it is dangerous if not kept under proper control. Many a promising young girl, or youth has been ruined by the subtle power that seeks to draw the innocent into the various avenues of vice. Its alluring, seductive fascinations operate like the wily

snake when attempting to charm an unsuspecting bird; and should be carefully avoided by all who wish to attain the character, and live the lives of true men and women.

2. That which is exerted principally through the channel of the intellect, controls those whom it affects, in thought and deed, in ways foreign to that which would be sanctioned by their interior and better sense of right. It is used in proselyting in behalf of the various shades of religious belief, the getting up of revivals, and engendering a disposition to wrangle in reference to the exciting political affairs of the day. Persons through whom this power is exercised in an intense degree, in appeals to a popular audience, can by argument, whether well founded or specious, carry the audience with them. They succeed for the time being, whether they are right or wrong. The influence abates after a time, or is counteracted by the recipient coming in contact with other minds who possess strong reasoning power, rightly directed. As we have intimated, if persons using this power are in the right —persons of noble souls, who would spurn a mean act—a great deal of good can be done with it; but if, on the contrary, it is exercised by the selfish and unprincipled, for evil purposes, much injury to society must necessarily be the result. Many cases might be cited, illustrating its deleterious effect upon susceptible individuals; as well as instances where persons have at first been repulsed by the antagonism of others who claimed respectability in regard to moral character, but after a few conversations, have yielded to the bewildering influence of false words

uttered with psychological power, and have become their admirers; held by that power until circumstances soon following revealed their real character, proving them to be guilty of base deception. If the victims had had decision of character, and followed their first impressions, feelings or intuitional promptings, they would have avoided the trouble and disgrace of their later experience, which is always sure to follow. Many marriages have been consummated under this power, wherein the attraction has lasted no longer than the duration of the power. Then the charm is replaced by inharmony and misery.

3. The third form is of more importance than either of the others, because of the wider range, and more exalted experience it gives to the impressional recipient subjected to it, when invoked by proper aspiration. Disembodied spirits, to most persons in the present state of spiritual progress invisible, are about all classes of individuals, whether they be high or low in unfoldment; each person attracting to himself what is congenial to his most cherished desires. As "like attracts like," so is it with these invisible influences. They are distinct intelligences, having a will and a purpose, be it upon a higher or lower plane; and as we are free agents, are attracted towards us, or repelled by us from our sphere of attraction. If we have no aspiration for noble ends, those of a lower plane, especially when they find us in a negative condition, will take possession and draw us into a still lower sphere of life than corresponds to our partial unfoldment. They enjoy the control of a person in that way as much as if in the material form. In

such cases, after a person has yielded to the influence which has been brought to bear upon him, it requires the efforts of a powerful magnetizer, or one who possesses strong psychological power to dispossess the deleterious influence, after which those of a better class will be attracted. It is an important fact that this unprogressed class can be reasoned with and improved in their condition, as much so as those who wear the covering of materiality.

The persecution which Theodore Parker suffered in earth-life, while earnestly endeavoring to improve the condition of his race, is an example of malignant influence, the offspring of intolerant bigotry. The following question was addressed to him in reference to it after he had entered spirit-life :

"During the latter part of your earthly life, prayer was instituted throughout the churches of a certain sectarian denomination, that God would either convert you or take you out of the world. Is it true that this concentration of many minds acting in unison for a special purpose did produce the desired effect. Was your health affected, or your death hastened, by these unhallowed prayers?

"Ans.—It is a well known scientific fact that the human body is to a very great extent, a psychological machine, because it is itself capable of being acted upon, either for good or ill, by all other minds. If a Jesus of Nazareth, or a magnetizer could restore a diseased body, giving health through the influence of psychological law, it is reasonable to suppose that the counter influence could as well be exerted, and with as much potency. During the last few months of my earthly life I clearly recognized the baneful psychological influence that had been exerted upon me from the source of which your question treats. As I neared the boundaries of the spirit-world it became more and more clear to me. I did not recognize it in the light of a wrong, but I recognized it as a power used in accordance with infinite law. As I was under the domain of law—not at any time exempt from it—I must bow before its decree. The forest trees fall before the storm. Who shall say it is not well? The lightning shivers the giant oak, and I believe it to be God's decree. Since I have become an inhabitant of the world where law is more understood than here, I have become clearly satisfied that my mortal life was shortened—perhaps many months—in consequence of this psychological influence. Their prayers were heard and answered. It was well. But as the great controlling influence of life makes use of all conditions for good, he made use of this. Every single phase of life, whether greater or lesser good, I believe to be in the hands or under the guidance of the great all-wise power of the universe—of all universes—and that what men act upon for seeming evil is always changed to good. The time is coming when these same persons who acted through their darkness will behold precisely where they stood, and how nearly they were related to justice, and how nearly I was related to justice—in what religious light we both stood. The scales will fall from every eye in due season, and every soul will be made to understand its relations to every other soul, and to the great God from whence we have come."

There can be no doubt of the existence of a high class of invisible intelligences, (invisible to all but clairvoyants) who respond to good desires, and in a variety of ways assist those who are in trouble or affliction; who are capable of sharing in the enjoyments which their aid confers, and of sympathizing with those who are in sorrow, and thus of mitigating the trials and afflictions which are incident to human life. All persons, whether they are willing to acknowledge it or not, are at times assisted by them both in sickness and health. There are as many kinds of these influences, when viewed in all their minutiæ, as there are differences in individuals; and each person attracts what his material body is chemically, and his spirit morally in a condition to receive.

It is advisable, indeed imperative upon us so to live as to be able to aid those who are below us in development, to a higher condition, and not allow ourselves to be drawn by degrading influences into a lower plane than our present moral status. Influences which our interior nature revolts at, whether exerted by persons in the form or out, we should courageously resist; and thus attract towards us those intelligences who will inspire us with noble thoughts, that will elevate our own souls, and reflect a benignant influence upon the community in which we live. If the latter influences, as well as the laws of life in the material and spiritual bodies were rightly understood and appreciated, we could now have a heaven on earth, as well as to wait until the change called "death."

Many of our public speakers of both sexes give utterance to thoughts that evidently come by impression or intuition, and are doubtless derived from this source of inspiration. The intelligences, although to them invisible and hence unrecognized, are as a real to others as any thing in material life. Many who have received these truths are willing to promulgate them, while others, equally convinced, lack the moral courage to make a public acknowledgement. Their reticence does not invalidate truth, which is eternal. Many promulgate the truth in reference to these things under other names. The change towards a belief in them is rapidly extending; and as we live in an age of progression, the fear of candid investigation will soon be dispelled; and the knowledge of the present life and of life as we learn of it in the spheres be spread abroad.

Having two bodies, the material and spiritual, the one as necessary as the other, as heretofore stated, the influences before mentioned affecting our spiritual condition affect our bodily health also, improving its condition when sought in conformity to natural law.

Belief is not always according to desire. The mind must receive as truth whatever proceeds from intuition, as well as what is proved by the evidence of the senses, and communicated to us, supported by credible testimony. Thus these views are submitted to the reader, not asking their acceptance, unless in accordance with his convictions of truth. We need not fear the truth, whether it emanates from an undeveloped source, or the spheres of higher unfoldment. Honesty in thought and deed should be our motto; and we should always be willing to receive the truth, whether it agrees or conflicts with our former cherished ideas, and educational teachings.

In earth life we often meet with persons who possess a beautiful spirit, but which is like a jewel inclosed in a poor setting, or as much obscured as a diamond embedded in the earth. Existing thus from birth, they live an antagonistic life, the spiritual aspirations being at war with the material tendencies much of the time. It may be considered a constitutional defect. Yet it is not irremediable. By the application of magnetism the material body can be spiritualized, so as to harmonize with the spirit. This is the only treatment known to be capable of reaching such cases. It can only be fully understood as a truth by those who know it from actual experience, or who have been unfolded into a condition favorable for its acceptance.

The influence of imagination upon health is a singular and interesting psychological phenomenon. Much of what is called disease is imaginary; hence there is truth in the old adage "Imagination will kill—imagination will cure." Many a patient has been told by his physician that he was dangerously ill, and having all confidence in the physician's knowledge of his case, he has considered the question settled, and believed himself to be dangerously sick. If he had been told that he was not very sick—that he must take courage, not allowing his mind to dwell upon his disease, and he would soon be well, the conditions would be so much more favorable that the prospect of cure would be vastly better.

Such is the power of the imagination that if a physician who has the entire confidence of his patient, should say to him "Your heart is affected," or "Your lungs are diseased," the invalid would be sure to feel an unnatural beating of the heart, or a severe pain in the lungs. It is better that the truth be told; but it is not necessary for him to know the extent of his disease, and how every organ is affected. The magnetizer does not require to know the extent of the disease, for he can without it work in confidence and trust, often effecing cures that would have been considered impossible, if he and the patient had realized that extent. Both of them would in the latter case have abandoned treatment as hopeless. Therefore the magnetizer is often more successful than those well informed in medical science.

The following statement, taken from a French publication, relates to a marked example of the great power of the mental faculty alluded to, in averting what would probably have proved a fatal issue.

"Alexandre Dumas published some time ago, in a daily Paris paper, a novel, in which the heroine, prosperous and happy, is assailed by consumption. All the slow and gradual symptoms were most naturally and touchingly described, and great interest was felt for the heroine.

"One day the Marquis Dalomieu called on him.

"'Dumas,' said he, 'have you composed the end of the story now being published in the———?'

"'Of course.'

"'Does the heroine die in the end?'

"'Of course; dies of consumption. After such symptoms as I have described how could she live?'

"'You must make her live. You must change the catastrophe.'

"'I cannot.'

"'Yes, you must; for on your heroine's life depends my daughter's.'

"'Your daughter's?'

"'Yes; she has all the various symptoms of consumption which you have described, and watches mournfully for every number of your novel, reading her own fate in your heroine's. Now, if you make your heroine live, my daughter, whose imagination has been deeply impressed, will live too.'

"'Come, a life to save is a temptation not to be resisted.'

"Dumas changed his last chapter. His heroine recovered and was happy.

"About five years afterward Dumas met the Marquis at a party.

"'Ah, Dumas,' he exclaimed, 'let me introduce you to my daughter; she owes her life to you. There she is.'

"'That fine handsome woman, who looks like Joanne d'Arc?'

"'Yes. She is married and has had four children.'

"'And my novel four editions,' said Dumas, 'so we we are quits.'"

To overcome the effect of the imagination generally requires considerable time; but by gradually operating upon the patient, saying but little upon the subject upon which his mind dwells, the magnetic power displaces the ideas which had fed the morbid fancy, and enables the mind to take up subjects which are new, and thus by this means its functions are restored to healthful activity, when all other modes of treatment have failed. With this, as with other forms of disease, the effect is produced by simply changing the chemical magnetic life force, and that by virtue of a law which when understood, seems as rational as it is immutable.

The question has sometimes arisen, "Is unbelief a hindrance to cure?" In olden times the fact that at certain periods, and in particular cases expected cures were not performed *because of unbelief*, was prominently set forth. Experience shows that it has certainly something to do with the success of magnetizers in our day. When a patient secures the services of a magnetizer whom he or she has been anxious to employ for the purpose, the treatment is more successful than if one after another is employed, and the various magnetic influences promiscously applied, without confidence in the adaptability of any one of them. Some magnetizers have the tact, or magnetic attraction to establish confidence in the mind of the patient at once: with others it is difficult; and there are those who in certain cases can never gain it.

The remarkable effect of confidence is observed in the common experience of physicians. If there is an aversion on the part of the patient, the physician, it is quite evident, is employing his services under the greatest possible disadvantage. The action of certain functions has been arrested by disappointment in failing to get the services of the favorite physician, and their activity resumed on subsequently securing them. Cures however, are often performed on persons who have been entirely skeptical; indicating that there is an absolute power independent of faith, notwithstanding the great influence of this mental action over nervous persons.

As with the exercise of imagination, in increasing or diminishing the intensity of disease, so is it with the power of the will. Patients affected as those described in another section concerning nurses, are in a negative condition. If they could stimulate the positive side of their being, they might be enabled to ward off much of their own disease during its inception. We have known persons who were positive in this respect, but whose organization was not of a healthy character. Their will power however was remarkable. They were determined to overcome the disease, and by their force of will did so; the patient in one case continuing to live twenty years after physicians had pronounced the prospect to be hopeless. Those who banish their disease from their thoughts, putting themselves in a condition to bury it in forgetfulness, or exercise their force of will as in these cases, will recover much sooner than if they permit their minds to dwell upon it, and give up their courage immediately upon suffering the slightest painful or disagreeable sensation. When they become discouraged, indicating entire passivity by word and act, they soon become so feeble as apparently to lose the power of utterance, drawling out their words with feeble listlessness. If they would arouse themselves, and try to speak with ordinary energy, in a natural tone, they would attract the positive healing influences, which assist nature to throw off the disease. This effort must come spontaneously from the patient, and not be compelled by the suggestions of others, however urgent or well-meant.

It is right to remain positive to whatever we know to be injurious, either physically, mentally or morally; and negative to the good and true, which will lift us into a higher condition of life, both in the material and spiritual realms. There are individuals so self-possessed, that they can go among persons of all conditions of life, and may be surrounded by a variety of influences, yet remain uncontaminated by whatever is evil in its nature; but on the contrary can assist and benefit those who are on a lower plane, with whom they come in contact, bringing them to a higher condition of unfoldment. Others again do not possess the positive power to resist injurious influences; and such are liable to be drawn down into a lower condition of life by associating with undeveloped persons whether in the material body or out. Hence it requires caution and care on the part of sensitive, negative persons, to avoid mingling with such as are calculated to affect them injuriously; and whose influence can be of no particular benefit to any. To the positive, there is little or no danger; but those of the opposite class are so numerous that a word of caution is not out of place. We must deal with persons as we find them, and not conceal that which should be extracted by the roots.

It is evident then, from the foregoing considerations, that the mind has much to do with the physical body. When it is calm and undisturbed, the body is in a better condition to resist the encroachment of disease. Persons should learn self-control, adapting themselves to the circumstances which surround them. If these are not as they would like them to be, they should still exercise patience and employ all reasonable means to improve their condition, embracing the first proper opportunity which offers a change for the better, submitting in the mean time without murmur to whatever in the nature of things is unavoidable. Thus we may know that if we do our part nobly, all things will work for the best in the end.

If the mind is too strongly exercised in one direction, and its powers concentrated upon one object, it will become unbalanced. When one faculty is steadily used to the exclusion of all the others, the person will become a monomaniac. We should not set up for ourselves idols to worship, nor indulge pet ideas; but cultivate all our powers equally, or as they may require, to maintain them in equilibrium, in order that our whole moral being may be harmoniously unfolded, and brought to its highest estate.

In illustration of the power of mind over disease, we may mention that we have known persons sick and lame, who on concentrating their minds upon a magnetizer, in connection with their malady, and coming in rapport with him, even without physical contact, were restored to health. Much can be done, as heretofore stated, to ward off disease by the exercise of the mind or will power; and it may even be eradicated after being fully formed, and reaching the chronic stage.

Over-taxing the mind is vastly more injurious and dangerous than inordinate physical exercise, as the mind controls the body. In sickness of any kind, if the mind is at rest, and the surroundings of the patient are pleasant, causing a tranquil feeling, the recovery is more rapid. There is on this account, as much benefit in good nursing as in the services of a physician, and more in the state of the patient's mind than either; therefore the benefit of cheerful influences should always be secured.

By the exercise of the will power the moral conduct can in a great degree be regulated and controlled. We frequently hear the epithet "unprincipled" applied to the selfish, whose conduct is not in accordance with the golden rule—that rule which was enjoined by Confucius, Jesus and others, and justly called *golden* because of its containing the very essence of a grand moral principle, worthy of universal obedience. Nevertheless, in all professions, in all ranks of society, and among persons of every religious belief we meet with men and women who disregard it. All flocks have their black sheep. We cannot therefore engage in any branch of business, or become members of any sect or society without mingling with both "good" and "bad," as the two classes of persons are termed respectively, according to custom. The difference between those so denominated is that goodness is undeveloped in some, while it is more or less developed in others. None are self created: all are creatures of circumstance. But if they are strong of will they can change their conditions. By this effort to improve, they can receive aid also from others. Reason is given us to distinguish right from wrong, and after getting to know the right, we should be courageous enough to pursue it, living as reasoning beings, and not allow ourselves to be drawn down to the plane of brute life. The same power that controls the higher developments of life controls the lower. A gradual change of development is going on among persons in all conditions of life, from the lowest to the highest. They have attained to as much good as as they can in their own sphere of unfoldment; but every one can be made better as fast as he progresses out of bad habits, and rectifies erroneous views of the duties and relationships of life.

Change also takes place unavoidably in matters of opinion. Nothing in life is fixed and unchangeable. Mutation is marked upon the face of every thing. As the objects in Nature change, in the progress of decay and renovation, so is it with forms of faith. To know and appreciate truth, we must be willing to investigate all things, following the Apostle's injunction, "Prove all things; hold fast that which is good." Separate the gold from the dross; compare the past with the present; lay aside prejudice, and accept truth, come from what source soever it may. We need not give up our individuality, trusting to others to think for us; neither should we require others to think as we do. Unless the evidence is convincing to their judgment, they cannot yield an understanding assent; for, as Goldsmith says, to convince another against his will, leaves him of the same opinion still. Pretended belief is worse than denial. It is better to be honest in the avowal of one's opinions and belief, let them be what they may, than to practise hypocrisy.

But while individual opinions often change, the laws

of the Universe continue in operation from age to age the same. The facts which have been elicited with reference to the practice of magnetism do not depend upon our belief. The practice rests either upon a true or a false basis; and there is no midway of doubt or uncertainty. We may see things to-morrow differently from what we do to-day; but facts are absolute and indisputable. If there is an apparent difference the fault lies in us and not in the fact.

A singular psychological phenomenon was recently observed to occur in the city of Boston. Two young men born of respectable American parents, being now of the ages of eighteen and twenty years respectively, from birth subject to an invisible power, in such a manner as to speak in foreign languages, which no one in their vicinity could understand. They were both affected alike, and conversed freely together, apparently understanding each other with facility. At the same time they could not speak so as to be understood in their own language. Children both older and younger in the same family are intelligent, but exhibit none of the peculiarities of clairvoyance, nor seem in any degree susceptible to psychological influence. In the case of the two brothers, we attribute the cause to ante-natal influences produced through impressions made upon the nervous system of the mother.

The case of the celebrated Mr. Home affords a still more remarkable example of extraordinary psychological control. We cannot enter into details of the very striking features of the phenomena which have been developed in his case; for they are still fresh in the public mind, the recent testimony of two English noblemen having again excited curiosity and astonishment in the minds of those not already familiar with them. The phenomena have also recently acquired sufficient importance to be assigned to a scientific committee for investigation. By telegram from London it is stated that the results of their yet incomplete examination are recorded in the scientific journals of that city; and further that the investigators are satisfied of the immense scientific importance of the subject. Professor Crookes and Sergeant Cox both seem to be convinced of the existence of a "nerve atmosphere of various intensity enveloping the human structure." A rational explanation of Mr. Home's experiences, upon a scientific basis, will be a source of great satisfaction to the inquiring mind.

In the Atlantic Monthly for August 1868 an article appeared entitled "Wonderful physical manifestations," giving an account of strange occurrences which took place in Fitchburg, Massachusetts, a short time previous. The case is altogether a remarkable one. We barely allude to some of its features, referring the reader to the original paper for a more complete account. The subject was a young Irish girl, who had obtained employment in the domestic service of a family in that town. Soon after she had been domiciled there, physical manifestations of an astounding character began to take place: a soapstone from the sink, weighing some twenty pounds was brought into the room where a portion of the family were seated, without physical contact, and laid at the feet of her employer; he at the same time closely watching her, to detect any attempt at deception; but of this there was not the least indication. Chairs were moved without contact, and many things equally strange were witnessed. Some of the occurrences seemed to point to the fact of obsession. The persons cognizant of the facts not understanding the psychological nature of the phenomena, the girl was placed in a Lunatic Hospital, from which she recently passed to the spirit world. One circumstance here to be observed is, that the manifestations did not occur at any time when the subject through whom, or in whose immediate presence they were produced, was alone; nor were they resumed after her removal from her temporary home in Fitchburg; showing that it required certain surrounding conditions to produce them; and that to a certain extent the presence of those who had been identified with them was essential. Undoubtedly if the precise quality of the nerve aura surrounding this case had been inquired into with a scientific purpose, as in the experiments with reference to Mr. Home, there would have been no need of the girl's incarceration, and in all probability such an unfortunate abridgement of her earthly career would not have occurred.

We may in this connection, before concluding the section on psychological phenomena, state the difference which exists between Mesmerism and Magnetism. A historical sketch will be found in subsequent pages. In the time of Mesmer but little thought was bestowed upon the uses of magnetism, beyond the effect observed on the functions of the brain after making passes upon the head. The mental powers of the subject were found to be controllable by this process, and made subservient to the will of the operator; and to this phenomenon the term *mesmerism* was given. It was applied to no practical purpose: indeed until within the past twenty five years it was supposed by the magnetizer that the effects produced by experimenting upon the person who had consented to be used as a subject for the purpose, were due to his own individual power, rather than to the magnetic force that operated through his material organism, independently of his spiritual or strictly personal identity. Subsequently it began to be perceived that the subjects passed beyond the control of the mesmerizers, who no longer possessed the power to release them from what had theretofore been often called mesmeric sleep, as they had done in the beginning of their experiments. They at length learned from the new phenomena which came unexpectedly before them, that their subjects had taken a higher form of development; and those who were zealous enough in the service of science to pursue the investigation of these phenomena to their final issue, or as near to that result as they were then able to advance, were rewarded with a more comprehensive and accurate view of man's psychological relations. It has been admitted that in magnetizing for the purpose of eradicating disease, it is not necessary to manipulate and infuse the operator's magnetism to such an extent as to control the mental faculties; neither is it judicious to bring under complete control those who are not so far prostrated by sickness as to be incapable of taking care of themselves.

In the practice of the magnetic treatment the patient can be aroused when sluggish, or quieted when excited, without affecting the brain in the way of sleep, or by bringing them into the trance condition. There is no need of this unless the patient has been deprived of sleep, and then this state of repose can be induced naturally, by the harmonious tendency of this method of treatment. It should not be attempted except in cases of insanity, and conditions approaching it. Experiments, however, for the purpose of showing the power of mind over mind are not improper.

Mesmerism and psychological power are the same. Magnetizing for the purpose of eradicating disease is altogether different. When the patient's system is fully charged with vital magnetism, it assists nature to regain the healthful exercise of the vital functions.

CONTRAST BETWEEN MEDICINE AND MAGNETISM.

We have thought a few words appropriate by way of contrast between the ordinary practice of medicine and magnetic treatment.

In the first place, it is contended by some that it is necessary for every one who practises medicine to have a diploma from a Medical College; and that without it they are not fitted for, or capable of relieving the sick and suffering; it matters not who is the person, or whether he is really qualified or not. Many diplomas are without doubt bought with money, when the holders have not possessed the ability, or have not had the experience which is necessary to guarantee qualification. We think no one will doubt this, who has investigated the subject.

A true physician adopts all means which he believes will afford relief to his patient, whether sanctioned by books and institutions or not. There are cartain persons known as bone-setters, who from birth almost, seem endowed with natural ability for that purpose. They get their diplomas from the highest source—the Author of their being, which will stand the test of all institutions combined. They could not follow any other business or profession successfully, but are widely known throughout the country for their success in this. Acquired knowledge is well, but all magnetizers claim that without natural ability no one is fitted to practise medicine. Both should be combined: and if to these are added the qualities of a powerful magnetizer, success is sure to follow their efforts.

We do not claim that medicine is useless in this age, in the hands of certain persons; but we claim that vital spirit magnetism is vastly superior to it, and to all systems of treatment by other means. A writer truly asks "What great advantage is it to be able to explain to a patient the nature of his disease; to discuss with more or less skill the origin, the seat, the progress, the probable consequences of the affection from which he suffers, but to be unable to relieve him?"

In many cases that have been treated by the ordinary practice, it is more difficult to eradicate the effects of medicine than it would have been to remove the disease, as it often happens that improper medicine develops new forms of disease worse than those treated.

In several States laws have been passed, making it a penal offence, punishable by imprisonment in a penitentiary or State Prison, to practise the healing art without a diploma from some established medical college.

In one State the law has been repealed, and others have refused to pass such a law. The effort to procure legislative aid and protection we regard as evidence of weakness. We contend for the right of all to choose for themselves whom they will employ for advice and treatment when sickness overtakes them. The following extreme language was used by a critic to show his estimate of the practice of medicine in former times. "——For nothing equals, if we believe you, the frightful evils which were inflicted upon mankind by former medicine; this fatal art, as you call it, which for centuries has enjoyed the power to decide arbitrarily on life and death; which destroys ten times more than do the most murderous wars, and which makes millions of others infinitely greater sufferers than they were at first."

It is an acknowledged fact that diseases and the mode of treating them are all the time changing; that remedies which will relieve a patient now would not do so in the past, neither will they in the future; that what will effect a cure when given to one patient, will prove an injury to another, afflicted with the same disease. It follows, therefore, that those who can clearly see the condition of the patient, and who know the chemical qualities of the medicine required, must be the most successful. Intuition or clairvoyance would be of great assistance to all practitioners.

If these statements are true, it appears to us that the practice in many cases is experimenting with life in the material body. We suppose it can be safely said that one half the medicine given may be so considered; and hence it is that with the change referred to, new remedies are being yearly introduced.

Why is it that among physicians prejudice exists against the practice of vital magnetism? To our mind the reasons seem to be first, that it involves much more labor, and takes more time; hence it is less profitable. A physician will get as much for writing a prescription as a magnetizer would for spending from fifteen minutes to an hour with his patient. Then there is no opportunity to get a commission on prescriptions, which is a source of great profit to some. Again, it is not popular, nor according to the rule which they have laid down for themselves. Many physicians have removed beyond the narrow limits of these restrictions, and are proving themselves more effective in power than the tree which produced them.

Many distinguished physicians have left on record their views in relation to the use of medicine, expressing the opinion that the practice of medicine is inadequate to the needs of the public, and that as to the use of drugs, the less that is employed the better for the patient. M. Lugol, a distinguished Parisian practitioner says that we must resort to some *new modes* if we desire to be more successful. Dr. Benjamin Rush said "The art of healing is like an unroofed temple; uncovered at the top, and cracked at the foundation." Dr. Jamieson of Edinburgh said that "nine times out of ten our miscalled remedies are absolutely injurious to our patients, suffering under diseases of whose real character and cause we are most culpably ignorant." Dr. James Jackson of Boston, a highly educated physician, propounded the following questions. "Shall we ever have fixed laws? Shall we ever know, or must we ever be doomed to suspect or presume? Is *perhaps* to be our qualifying word forever? Do we know, for example, in how many cases such a treatment fails for the one time that it succeeds?" Dr. Ramage, member of the Royal College of Physicians in London said that "It cannot be denied that the present system of medicine is a burning reproach to its professors; if, indeed, a series of vague and uncertain incongruities deserve to be called by that name. How rarely do our medicines do good! How often do they make our patients really worse! I fearlessly assert that in most cases the sufferer would be safer without a physician than with one. I have seen enough of the mal-practice of my professional brethren to warrant the strong language I employ." Dr. Jacob Bigelow of Boston, formerly President of the Massachusetts Medical Society said, "It is with regret that we are obliged to acknowledge a third class, that of incurable diseases, which has been recognized in all ages as the disgrace of the medical profession. Medicine in regard to some of its most important objects, is still ineffectual speculation." The distinguished Magendie of Paris gave the following opinion. "I hesitate not to declare, no matter how sorely I shall wound our vanity, that so gross is our ignorance of the real nature of the physiological disorders called diseases, that it would perhaps be better to do nothing, and resign the complaint we are called upon to treat to the resources of *nature*, than to act as we are frequently compelled to do, without knowing the why and the wherefore of our conduct, and at obvious risk of hastening the end of the patient."

If these statements and opinions of some of the most distinguished medical men of both the old and new worlds are founded in truth, is it not incumbent upon physicians every where to investigate the life forces of nature, and the laws of mind; and avail themselves of all means, natural or acquired, that will relieve the afflicted who place implicit confidence in their reputed skill?

HYGIENIC SUGGESTIONS.

While we consider Vital Magnetism a great boon to the human family, we would not encourage our readers to be too enthusiastic in their expectations of what it can accomplish. We do not believe that all that is necessary to sustain the physical body is to be found in magnetism alone. On the contrary, each individual must make use of whatever he has found to agree with him, whether derived from the mineral, vegetable or animal kingdoms; and in all things pursue such a course as his own personal experience has proved to be conducive to general health. In order to live properly and secure the enjoyment of vigorous health, we must eat and drink of the food placed within our reach, that meets the real wants of our nature: as to kinds and quantity, there is a great diversity in the needs of different individuals. Indeed, the fact has forced itself upon common observation that scarcely any two are alike in this respect.

There are habits of a vitiating tendency to be avoided, such as the improper use of fermented liquors, and opium, attention to be given to bathing, pure air to be secured in dwellings and places of business; and many other things that are essential to the enjoyment of good health. These are quite as important in the way of prevention, as magnetism is for the cure of disease; for although we may have the good fortune to be able to secure the highest vitalizing power of the latter agency, the neglect of the laws of health being in antagonism with it, would undo all that had been gained by its use.

If the patient has a besetting sin or vice, or a habit of any kind that is detrimental to health, it is the duty of the physician to advise abstinence therefrom, either immediate or gradual. A physician who neglects such advice is unworthy of the confidence of his patient, or of professional respect. It is useless to prescribe for a patient if he is allowed to undo in an hour, what it has required a week for his physician to do. In the relation of physician and patient, candor and common sense advice is the best for both; and it is the duty of the patient to follow such advice.

When the stomach is in a healthy condition, the appetite craves what is best for it; but if it is morbid, it should not be gratified. The nurse should use good judgment, in addition to medical advice, in preparing food; carefully avoiding that which is known to be injurious. People of observation and mature experience generally know what kinds of food and drink agree with them in health. Some articles disagree with, and indeed are decidedly injurious to certain persons; while to others the same articles prove not only harmless, but beneficial. After ascertaining what viands are indigestible, and consequently unsuited to the particular constitution, the person should have decision enough to abstain from them at once. In sickness this becomes highly important, and more especially is abstemiousness necessary where the injurious consequences of a fixed habit affect the mental, moral or spiritual welfare. Reason enables us to judge of the fitness of things, and this is aided and confirmed by intuitional promptings. If human beings would trust their intuition as the lower animals do their instinct, there would be less sickness and suffering in the world. Certain animals when sick leave off eating and drinking, and seek out for themselves some particular herb, of which they partake, and are soon restored to their ordinary health. At the same time they shun the poisonous herbs, giving us a lesson which it would be well to heed. When the appetite is depraved, magnetism or the use of simple, harmless medicines, which every family ought to be acquainted with, is all that is necessary to afford relief in a majority of cases. Over-eating or drinking of such solid and fluid food as is good in quality should in like manner be refrained from during sickness, that the stomach may have proper rest, and be able to resume its functions.

Bathing is essential both for cleanliness and health; but like other things when misused, or not properly regulated, it may do harm. Although seemingly very simple, it requires the exercise of judgment to employ it to the best advantage. Some can use cold water and feel refreshed, while with others it would be injurious, producing a chilliness too great for immediate reaction, and congestion would be the result. The length of time for remaining in the bath, and the frequency of its repetition, differ with different persons. By too frequent bathing some lose their vital force, and debility follows. No rule can be laid down for the guidance of all; but each one should study his own physical constitution, learning what there is peculiar about it that requires a departure from general rules, keeping the body clean and the pores of the skin open. The kind of bath required, as well as the quantity of water and the frequency of its use, is not always the same, but differs with the same indivduals at different times, and under varying conditions. A good magnetic application by the rubbing process after the bath, is beneficial to keep up a vigorous circulation.

The Turkish Bath, and what is termed "Vacuum Cure" are no doubt beneficial in some cases; but the difficulty is in knowing what cases they are adapted to. If applied by parties who are intuitively or clairvoyantly gifted, so as to distinguish those which can receive benefit from those which would be injured by the use of such measures, instead of applying them indiscriminately to all who desire treatment, they would be capable of doing much more good to humanity. The magnetic treatment accomplishes what is claimed for these modes of treatment without the dangerous effects liable to be produced when employed in cases to which they are not adapted.

Intoxicating liquors should never be used as a beverage. They are unnatural stimulants, which blunt the sensibility, so that persons using them habitually require constantly increasing quantities to produce the same effect. They lose their natural life force, and the blood becomes deteriorated, and watery in consistence. Disease of various forms follows the inordinate use of every alcoholic drink. Rheumatism and dropsy are not uncommon consequences, and consumption often cuts short the earth life. We are of opinion that their use as a beverage was never intended; and the sooner they are discontinued, the better for the consumer, his country, and the world at large. Those who use such liquors to excess as a beverage are not safe and trust-worthy, and therefore should never be employed in places of responsibility. It is a shame to see officers of Government so muddled with the effect of liquors in the afternoon, day after day, as to be unable to take care of themselves; instead of keeping clear-headed, and in a fit condition to take charge of public business, which requires the clearest intellect. Conductors and engineers of Railroads, and officers in charge of Steamboats, who have many valuable lives in their care, should be temperate men. Both companies and individuals should be forbidden by law to employ persons to fill positions of such responsibility, who are habitually intemperate.

It is a well attested scientific fact that Tobacco is a poison. It is also a powerful stimulant and narcotic. Its effects upon persons of a nervous temperament are produced gradually; but like those of opium and intoxicating liquors, they are sure; and if indulged in to excess, will finally produce injurious consequences to the whole physical system. When applied to the surface of the body in health, it produces a blister. It will also heal or remove the virus of poison, or inflammation in certain cases, when properly applied. It has its uses, therefore, but its general use by the thousands who consume it is an abuse; and it was never intended to be indulged as a daily habit. Many of its habitual slaves are so filled with its noxious effluvia, that they become offensive to those whose taste has not already been vitiated. To sit beside them, even in open conveyances in the public streets, is nauseating. If they could see themselves as others see them, they would out of self respect, one would suppose, abandon the habit. When the abuse of liquors is joined with it, it is so offensive to those who are compelled to live in the same house with them, and to be brought in daily contact, that it might make an angel cry out against them. It is strange that men and women will indulge in habits which their better nature spurns, and partake of subtances which the lower orders of animals, with their protective instincts shun. We would advise those who are willing to try to break up the habit, to procure a small quantity of Gentian root, and use it as a substitute for spirituous liquors, tobacco or opium; and when the acquired appetite has been abated, discontinue its use. Smoking gives great offense to persons of delicate and unperverted sensibility. King James said of it, "It is a custom loathsome to the eye, hateful to the nose, harmful to the brain, dangerous to the lungs." John Josselyn says of tobacco, although himself a lover of it, that "immoderately taken it drieth the body, inflameth the blood, hurteth the brain, and weakeneth the eyes and sinnews."

There is another habit widely prevalent, which is a source of serious evil, being destructive of both physical and mental powers. We refer to the secret habit indulged by both males and females, and which causes a rapid waste of the material life. It destroys those who are the most innocent in all other respects, and those of the brightest intellect. If persisted in for any length of time, it inevitably produces a state resembling idiocy; frequently insanity and consumption are consequences following the practice at no remote period. Our Insane Asylums abound with the victims of this indiscretion, in the development of whose mental disease it operated as the sole cause. "Timely hints given in kindness often do more good than long sermons." Parents and guardians should act upon this, making it a point to warn the young persons whom they have in charge, of the dangerous tendencies of the "solitary vice." Many a bright and promising youth, and many a beautiful girl may be saved from destruction of body and mind by proper caution. We have known cases of imbecility to originate solely from the victims being addicted to this habit. Magnetism was the only remedy that produced any beneficial effect, after medication had been tried for years without success. It is almost impossible to restore the nervous system to its natural tone and vigor, after being addicted to any of these indulgences, except by vital magnetism. It is useless to employ a physician unless such habits are broken up, and an effort made to strike at the root of the disease, the patient trying to regain what has been partially destroyed. This done, the physician would be much more successful in his treatment.

By living a true life, in obedience to the laws of hygiene, the business of the physician would be in a great measure superceded, and the druggists' shops, which in cities are on every third corner of the streets, might be converted into grocery and provision stores.

ANIMAL MAGNETISM.

The following compilation, derived from a variety of sources, gives a condensed history of the origin and progress of Animal Magnetism, and the philosophy of its action.

The sympathy between one human being and another, and between man and Nature may be inexplicable, but it is none the less real; and it indicates a vast system of correspondences, in which every affection finds its counterpart in some object, and every object is the emblem and corporeal image of some intellectual, that is to say spiritual cause. Esteem, friendship, love—how do these by a mysterious sympathy seek out their proper objects? Who shall say there is no starry road by which one longing spirit may seek another, and along which if we only have love and faith, the weakest of us may fly to the aid of a suffering brother or sister?

MESMERISM: We have no single term which expresses adequately the nature of the phenomena generally treated under this head. In France, Somnambulism and Animal Magnetism have been very generally used, and in England Mesmerism. The facts however, are so multiform and as yet so ill digested, that it is hardly worth while occupying any serious amount of time with the discussion of mere terms. A proper history of mesmerism or animal magnetism would commence with an account of the natural sleep, and of somnambulism or sleep-walking. To save space, we must presume that the reader has some acquaintance with the speculations concerning these phenomena, and the points of interest they present to an intelligent inquirer. The *facts* are numerous and interesting, and may be consulted in any medical Cyclopedia, or in the works of Abercrombie. It would be in vain, however, to search for a satisfactory account of causes, or for a theory more than plausible in such productions, which all regard sleep in proportion to its depth, as a mere negation of activity.

Antiquity of Magnetism. The power generally known as animal magnetism or mesmerism, which has been regarded as a novelty, was exercised in remote antiquity, and was properly the chief art of the magician. Ample proof of this fact is given by Ennemoser. We shall borrow same of his quotations from a French writer. Magnetism, he says, was daily practised in the temples of Isis, Osiris and Serapis. In these temples the priests treated the sick and cured them, either by magnetic manipulations, or by other means producing somnambulism. We shall prefer turning our attention to such Egyptian monuments as present us with whole scenes of magnetic treatment.

Although these Egyptian hieroglyphics are regarded with great daring and boldness, yet much that is probable results, and the more so from the fact that all things in these monuments are not hieroglyphic. There are also purely historical paintings, which represent sacrifices, religious ceremonies, and other actions as well as things which refer to the natural history of animals, of plants and the stars.

It is usual to imagine that all Egyptian subjects were emblematical, when in fact they were not, for hieroglyphics must not be confounded with emblems. The former (caracteres hieroglyphiques) are symbolical representations of whole chains of ideas, which at a later time were condensed; the latter are representations of separate actions. The hieroglyphics, he further remarks, were probably at first whole figures, but as they occupied too much space, they were gradually abbreviated, and portions alone remained lines, from which it was impossible for strangers to discover the original meaning.

Among the emblems he includes the remarkable representation on a mummy case given by Montfaucon. Before a bed or table, on which lie the sick, stands a person in a brown garment, and with open eyes, and the dog's head of Anubis; his countenance is turned upon the sick person; his left hand is placed upon the breast, and the right is raised over the head of his patient, quite in the position of a magnetizer. At both ends of the bed stand two female figures, one with the right hand raised, the other with the left. The bed was supported by four feet which bear the Isis' head, hawk's head, dog's head, and a human head; the symbols of the four healing divinities, Isis, Osiris, Anubis, and Horus. Other hieroglyphics on a talisman, bearing similar representations are mentioned and upon other mummies, where standing figures touch the feet, the head, the sides or the thighs; and many other magnetic actions are represented; these are produced in Montfaucon, and in Denon's "Voyage d'Egypte." These scenes do not stand alone. Figures occur in the amulets or charms known as Abraxæi; all more or less manifesting acquaintance with magnetism. The priest with the dog's head or mask of Anubis occurs repeatedly, with his hands variously placed on the supposed patient. Some of these figures are given by Montfaucon. In one of them the masked figure places one hand on the feet, the other upon the head of the patient; in a second, one hand is laid upon the stomach, the other upon the head; in a third the hands are upon the loins; in a fourth the hands are placed upon the thighs, and the eyes of the operator fixed on the patient's countenance. All these representations were involved in mystery until magnetism was re-discovered. Our French authority, however, like all theorists, bends every thing to his purpose, and certain figures in Denon, which evidently represent the resurrection of the spirit, and not the magnetic awakening in the body, are explained on the same principle. Ennemoser also fully adopts his hypothesis, and concludes with this remark. "Thus we see, in various stages of recovery, that the patient gradually rises from the couch, a fact which, therefore, excludes the idea of a dead body," as if the awakening spirit should be regarded as a dead body!

SYMBOL OF THE HAND. These bronze fingers are fore-fingers. Is it that the Egyptians magnetized especially with this finger? Magnetic somnambulists often magnetize with the fore-finger alone, and in cases of cramp order it to be used. It is remarked on the authority of Tomasius that the position of these bronze hands is the same as that of the prelates and popes when they blessed the people, and as that in which the painters of all ages have been accustomed to represent the hand of our Saviour. Indeed the mysterious hand is not confined to the Egyptian antiquities, but it re-appears in the coronation ceremonies of Europe, and after a time we begin to recognize it as a symbol of the royal gift of healing by touch. This, however, is not understood under its earlier forms described by Montfaucon. A hand, for example, is represented as descending from heaven, in a picture of Charlemagne, and in two portraits of Charles the Bald, Justice pointing with four fingers towards his head, to illuminate him in his duties and towards his subjects. From the fingers of these hands proceed rays. On a monument of Dagobert at St. Denis, a similar hand was represented, with three fingers extended, while the king naked, with a crown on his head, was raised over some drapery by two bishops, with two angels near him.

According to Montfaucon similar hands are common to the emperors of Constantinople about the period of

Charlemagne. From these and many similar documents of antiquity, Ennemoser is inclined to assign a divine origin to this symbol; in short, to recognize it as the hand of the Lord, so often named in the scriptures.

POWER OF THE EYE. It is then an actual power that we ascribe to the hand, without which it could never have become the symbol of power among the wise ancients. The power of the eye is equally remarkable, and even savage animals turn away from its fixed, dauntless expression. The fascination of the eye has been an article of the popular creed in all ages. All the passions seem to find a more instant manifestation in its mild or flashing light. Aubrey speaks of it with his usual felicitous simplicity, "Amor ex oculo—love is from the eye, but as Lord Bacon saith, more by glances than by full gazings. The glances of envy and malice do show also subtlety; the eye of the malicious person does really infect and make sick the spirit of the others." Children, he remarks, are very sensible of these irradiations. Wierius has an observation upon the abundant flow of spirits from the pupil of the eye, and its lightning-like glances. Every one must have felt how the soul, with fuller meaning and more intense purpose, rushes to the eye and kindles it with its most subtle fire. Magnetism by the eye is indeed often more powerful than by the hands, but there is probably a specific difference which experience may determine accurately. Comparatively few can exercise this power continuously, intently, and at the same time dispassionately, and none ought to attempt the use of it who are not master of their own purpose, and shall we say, pure in heart?

POWER OF WORDS, NUMBERS AND SIGNS. The will, after all, is the real power exercised by the magnetist, and consequently his influence must be good or evil, according to the ruling motive of his life. The hand, the eye, the expression, are not the power, but they give direction to it variously. The direction may be determined also by words, numbers and signs, or by the silent will of the operator, all which have a subtle or magical influence upon the patient. A bare intimation of the fact is all that our space will allow.

RELATION BETWEEN THE ANCIENT AND MODERN PRACTICE OF MAGNETISM. But the antiquity of magnetism is not an isolated fact; its existence gives form and color to the whole ancient world, and divides it from the modern with as much distinctness as the life of the dreamer or seer is separated from that of our day-light occupations. Ennemoser therefore justly observes that "Christianity was a very important crisis in the history of magic, in fact the most important." His remarks have so much suggestive value that it would be inexcusable to omit a larger quotation.

The advent of Christ is, in a historical point of view, the central era, when the old time comes to an end, and the new commences; when the night-like shadowiness of mysteries is dissolved into the day-light of self-consciousness, and the purpose and intention of life.

As the Biblical history of the Old Testament is the seed and the type of all later history, so in the New

Testament for the first time, like the flower unfolding from the bud, is developed a perfect revelation of the truth. The Judaism of the Old Testament involves in it a real perception of the true tree of life, of the inner progressive development by means of cultivation; all other heathen nations, with their various systems of religion, are the lopped branches of the great tree of life, which have vegetated, it is true, but which are incapable of inner growth. Judaism is that real mystery which appears in Christianity as the ideal of holiness, and union with God. But as the fruit is matured from the blossom only by progressive degrees, so too does this maturity in the new history advance forward with a measured step. Religion and morals, art and science, are it is true, progressing in new and widely ramifying parties in this latter christian time, but they are as yet very far from their goal, which is perfection.

The same may be said with regard to magnetism, which has yet advanced only so far as the intelligence of those minds which have labored to comprehend it, has *itself* advanced. The same thought is so aptly expressed in an old periodical, that we should be inclined to ascribe it to the same original as Ennemoser's.

"The ancient world" says the anonymous writer we allude to, "presents the phenomena of telluric influence. It is the night of mankind: here wonders, divinations, dreams, prophecies, oracles and revelations follow one another. As the animal by instinct builds the most wonderful nests or cells—moves and travels from region to region—distinguishes the healing plant or nourishing food from that which is poisonous and unwholesome:

in the same way the seers, the magicians, the priests, the poets, the artists of the old world performed those deeds which the most enlightened among the children living under the solar life can now neither understand nor believe. Thus the working individual can scarcely comprehend and believe that which he has dreamed or done during that part of life which he calls sleep.

The ancient world had reached the summit of telluric life, when Christ and the christian had made their appearance. The earth was then on the highest point of somnambulism. By means of Christ and the christian the family of mankind began to awaken. Christ's wonders are as it were, the morning dreams of one who is near to open his eyes to the beams of that centre of life which calls forth the solar light in nature.

For the principle involved in this distinction, recourse must be had to Swedenborg, who establishes that a radical change in the mode of perception has taken place since the most ancient times. In the study of the subject, we place our finger upon the key to all human history and philosophy.

THE EARLY CHRISTIAN PERIOD. The "Acts of the Apostles" contain evidence that the ancient practice was now christianized, not absolutely abandoned, and much valuable evidence concerning the use of this occult power may be collected from the writings of the fathers. Vision and prophecy, understood as directly appointed, ceased with Malachi, but it commences again when the angel appears to Zacharias and the Virgin Mary. The experimental knowledge of ecstasy and spiritual influence originated many remarkable practices among the primitive christians. (See the writings of Augustine, and Ambrose.)

The pagan temples were succeeded by the christian monasteries, and in these last receptacles the divinatory faculty found a second sanctuary, a refuge and a home.

The gift of divination being natural to the species, under certain conditions of its development and exercise, is still continued to be manifested among the converts to the new faith; and although under a somewhat altered form, it was still enlisted as previously into the service of the priesthood, and devoted to the purpose of religious worship. Although the christian sybils and pythonesses no longer sat upon the tripod—although they ceased to utter their predictions in public, their prophetic faculty still accompanied them wherever they went.

THE MIDDLE AGES. The power of magnetism, either theoretically or practically, was never wholly unknown. In Asia or China it has probably never ceased to be practised, from the remotest antiquity down to the present hour, and in the former vast region of population its use has been varied by that of drugs and narcotics. The writings of Avicenna and the annals of the Roman Catholic worthies may be consulted; and in English literature, the ecclesiastical history of Bede, who has placed on record many remarkable cures, performed some ages before, both by the hand and by prayer. In Bede's time there was little question of philosophy, and it was four or five centuries later before the Universities arose. The occult sciences participated in the revival of learning, and the middle age period of magnetism, dreamy and profitless for many good reasons, closes with several great names, e. g. Paracelsus, Cornelius, Agrippa and others.

MESMER. The man fated to produce this cloistered wisdom of the elder times upon the busy stage of life was a Swiss German, by name Frederick Anthony Mesmer, whose birth dates in 1734. "In his earliest youth" says one of his biographers, "he evinced a great fondness for the study of nature; he told me that when a boy, his greatest pleasure was to retire into solitary spots, and there to amuse himself in contemplating the operations of insects, the flight of birds, and in comparing the different shapes of plants, herbs and mosses. He remained often out in the fields till late in the night, when the rising of the stars and the moon filled him with deep sacred feelings. 'I was then' said he, 'under the magnetical influence of nature; the full flood was streaming above, below, and around me. My mind was full, but I did not know what was working in me.'"

As a student of medicine he showed great independence of thought, and his favorite books were the almost forgotten labors of mystics and astrologers. On obtaining his degree in 1766 he published his inaugural dissertation De Planetarum Influxa (On the influence of the Planets)) which marked him out among his professional brethren as a confirmed visionary. His theory supposed the magnetic element to pervade the entire Universe, and penetrate all bodies, acting in the same relation to the nervous system of all animals as light to the eye.

Settled in the Austrian capitol as a physician, he occupied his leisure hours with the attempt to bring this theory into practice as a means of cure, and at last, towards the close of 1773, resorted to the artificial magnet, his coadjutor in these experiments being the Jesuit Maximilian, Hellenic Professor of Astronomy at Vienna. The latter afterwards claimed the priority of discovery, which produced some disagreement between them, and probably had some effect in turning Mesmer from the use of artificial means to the more exclusive study of "Animal Magnetism," as he finally termed it.

Somnambulism also discovered itself to him while he treated some of his patients with the lode stone, and he may now have surmised that all the divine virtues attributed to its magnetic properties by the ancients were capable of realization. It is not necessary, when all that we can say on this subject must be confined to the narrowest limits, to follow Mesmer step by step in his discovery. In 1775 he found it convenient to quit Vienna, and occupied that and the following year in traveling through Bavaria and Switzerland, where he effected some remarkable cures, both in private circles and in the public hospitals. On returning home he established a hospital in his own house, but stood in such ill repute that he was ordered to quit Vienna, and in the beginning of 1778 he sought a new theatre for the exercise of his art in Paris. Here Bergasse and Dr. Eslon became his ardent disciples, and Mesmer, whose character was not without its weak points, assumed the airs of a magician, with a view to secresy, and perhaps to greater gain. Encouraged by the later of his converts, he published in 1779 his first treatise on Animal Magnetism, and his imperfectly developed theory is thus stated and commented on by his personal acquaintance to whom we have already referred.

There is a reciprocal influence (action and re-action) between the planets, the earth, and animated nature. The means of operating this action and re-action is a most fine subtle fluid, which penetrates every thing, and is capable of receiving and communicating all kinds of impressions. This is brought about by mechanical, but as yet unknown laws. [It may be so at that period of time.] The reciprocal effects are analagous to the ebb and flow. The properties of matter and organization depend upon reciprocal action. This fluid exercises an immediate action on the nerves, with which it embodies itself, and produces in the human body phenomena similar to those of the lode-stone's, that is polarity and inclination. [Here was a great mistake of Mesmer, of confounding the original law of polarity and life with the effect of a particular fluid.] Hence the name of animal magnetism. This fluid flows with the greatest quickness from body to body, acts at a distance, and is reflected by the mirror like light; and it is strengthened and propagated by sound.

There are animated bodies which exercise an action directly opposite to animal magnetism. Their presence alone is capable of destroying the effect of magnetism.

This power is also a positive power. [How can there be two positives, one opposite to the other?]

By means of animal magnetism we can effect an immediate cure of the nervous diseases, and a mediate cure of all other disorders; indeed, it explains the action of the medicaments, and operates the crisis. The physician can discover by magnetism the nature of the most complicated diseases.

We have allowed the fanciful words "telluric" and "solar" to stand without adopting them, our belief being that we are even now, a quarter of a century since it was written, far from being in a position to select our final terms, and express the magnetic doctrine in proper form. Such however, were the announcements of Mesmer, and the subsequent comments on them.

The scenes around his magnetic battery, in the meanwhile had drawn the attention of the French Government to his proceedings, and in 1784 the first commission was appointed to examine them. The members consisted of four physicians, one of whom was the luckless Dr. Guillotine, and five members of the Academy, Franklin, Leroi, Bailly, De Bory, and Lavoisier. The report, drawn up by Bailly was unfavorable to the truth of animal magnetism, and in conclusion, it denounced the pernicious tendency of the practice, as well physically as in morals. But the whirl of the French revolution was now just commencing, and magnetism and clairvoyance were presently associated with political and social aims.

A grand Societe de l'Harmonie was formed on the principle of the Harmonic Circles in America at the present time. Large sums of money were subscribed, by which Mesmer enriched himself, and the famous Marquis Puysegar became an adept. In fine, Mesmer was obliged to quit France, and after residing some time in England under a feigned name, he died in his native place, assured of the unobtrusive progress of his doctrines, in 1815.

RECENT HISTORY OF MESMERISM. In 1813 M. Deleuze of Paris had written his well-known "History of Animal Magnetism," and the turmoil in which this discovery had been involved by the revolution having subsided, the subject was again open to sober observation. It had also, within the last twenty years made the tour of Europe, and some of the most illustrious *savants* and men of letters in Germany had addressed themselves to the investigation. Several distinct schools began to appear in France, the most important of which was under the direction of Puysegar at Strasburg. The battery and its crisis were dispensed with, and the intelligible observation of psychological phenomena was now connected with the physical treatment.

In 1825 Deleuze published his Practical Instructions, and about this period such was the urgency of the revived interest in this subject, that it was brought under the observation of the Royal Academy of Medicine.

In February 1826 a new commission was appointed, whose labors extended over five years till 1831, when their report was drawn up; it was signed by Bourdois de la Motte, President, and MM. Fouquier, Gueneau de Mussy, Guersont, Hufson, (who drew up the report.)

Viewing this report as the turning point in the modern history of magnetism, for it was favorable to the practice, we shall insert its "conclusions," the text itself being too voluminous.

The commission resolved as follows: 1. The contact of the thumbs or of the hands, frictions, or certain gestures, which are made at a small distance from the body, and are called *passes* are the means employed to place ourselves in magnetic connection, or in other words, to transmit the magnetic influence to the patient.

2. The means which are external and visible are not always necessary, since on many occasions the will, the fixed look, have been found sufficient to produce the magnetic phenomena, even without the knowledge of the patient.

3. Magnetism has taken effect upon patients of different sexes and ages.

4. The time required for transmitting the magnetic influence with effect has varied from half an hour to a minute.

5. In general, magnetism does not act upon persons in a sound state of health.

6. Neither does it act upon all sick persons.

7. Sometimes, during the process of magnetizing, there are manifested insignificant and evanescent effects, which cannot be attributed to the magnetism alone; such as a slight degree of oppression, of heat or cold, and some other nervous phenomena, which can be explained without the intervention of a particular agent, upon the principle of hope or fear, prejudice, and the novelty of treatment, the *ennui* produced by the monotony of the gestures, and the silence and repose in which the experiments are made, finally, by the imagination, which has so much influence on some minds, and on certain organizations.

8. A certain number of the effects observed appeared to us to depend upon magnetism alone, and were never produced without its application. These are all well observed physiological and therapeutic phenomena.

9. The real effects produced by magnetism are very various. It agitates some and soothes others. Most commonly it occasions a momentary acceleration of the respiration and of the circulation, fugitive fibrillary convulsive motions resembling electric shocks, a numbness in a greater or less degree, heaviness, somnolency, and in a small number of cases that which the magnetizers call somnambulism.

10. The existence of uniform character, to enable us to recognize in every case, the reality of the state of somnambulism has not been established.

11. However, we may conclude with certainty that this state exists, when it gives rise to the development of new faculties, which have been designated by the names of clairvoyance, intuition, external prevision; or when it produces great changes in the physical economy, such as insensibility, a sudden and considerable increase of strength; and when these effects cannot he referred to any other cause.

12. As among the effects attributed to somnambulism, there are some which may be feigned. Somnambulism itself may be feigned, and furnish to quackery the means of deception. Thus in the observation of those phenomena which do not present themselves again but as isolated facts, it is only by means of the most attentive scrutiny, the most rigid precautions, and numerous and varied experiments that we can escape illusion.

13. Sleep produced with more or less promptitude is a real, but not a constant effect of magnetism.

14. We hold it as demonstrated that it has been produced in circumstances in which the persons magnetized could not see or were ignorant of the means employed to occasion it.

15. When a person has once been made to fall into the magnetic sleep, it is not always necessary to have recourse to contact, in order to magnetize him anew. The look of the magnetizer, his volition alone, possess the same influence. We can not only act upon the magnetized person, but even place him in a complete state of somnambulism, and bring him out of it without his knowledge, out of his sight, at a certain distance, and with doors intervening.

16. In general, changes, more or less remarkable, are produced upon the perception and other mental faculties of those individuals who fall into somnambulism in consequence of magnetism. Some persons amidst the noise of a confused conversation, hear only the voice of their magnetizer; several answer precisely the questions he puts to them, or which are addressed to them by those individuals with whom they have been placed in magnetic connection; others carry on conversation with all the persons around them. Nevertheless, it is seldom that they hear what is passing around them.

During the greater part of the time, they are completely strangers to the continual and unexpected noise which is made close to their ears, such as the sound of copper vessels struck briskly near them, the fall of a piece of furniture, &c. The eyes are closed, the eye-lids yield with difficulty to the efforts which are made to open them; this operation, which is not without pain, shows the ball of the eye convulsed, and carried upwards, and sometimes towards the lower part of the orbit. Sometimes the power of smelling appears to be annihilated. They may be made to inhale muriatic acid or ammonia, without feeling any inconvenience, nay, without perceiving it. The contrary takes place in certain cases, and they retain the sense of smelling.

The greater number of the somnambulists whom we have seen, were completely insensible. We might tickle their feet, their nostrils, and the angle of their eyes with a feather; we might pinch their skin so as to leave a mark, or prick them with pins under the nails, without producing any pain—without even their perceiving it. Finally, we saw one who was insensible to one of the most painful operations in Surgery, and who did not manifest the slightest emotion in her countenance, her pulse, or her respiration.

17. Magnetism is as intense, and as speedily felt at a distance of six feet as of six inches, and the phenomena developed are the same in both cases.

18. The action at a distance does not appear capable of being exerted with success, excepting upon individuals who have been already magnetized.

19. We only saw one person who fell into somnambulism upon being magnetized for the first time. Sometimes somnambulism was not manifested until the eighth or tenth sitting.

20. We have invariably seen the ordinary sleep, which is the repose of the organs of sense, of the intel-

lectual faculties, and the voluntary motions precede and terminate the state of somnambulism.

21. While in the state of somnambulism, the patients whom we have observed, retained the use of the faculties which they possessed when awake. Even their memory appeared to be more faithful and more extensive, because they remembered every thing that passed at the time, and every time they were placed in the state of somnambulism.

22. Upon awakening, they said they had totally forgotten the circumstances which took place during the somnambulism, and never recollected them. For this fact we can trace no other authority than their own declarations.

23. The muscular powers of somnambulists are sometimes benumbed and paralyzed. At other times their motions are constrained, and the somnambulists walk or totter about like drunken men, sometimes avoiding, and sometimes not avoiding the obstacles which may happen to be in their way.

There are some somnambulists who preserve entire the power of motion; there are even some who display more strength and agility than in their waking state.

24. We have seen two somnambulists who distinguished, with their eyes closed, the objects which were placed before them; they mentioned the color and value of cards without touching them; they read words traced with the hands, as also some lines of books opened at random.

It is much easier to deny such facts than to account for them, but instances like this are too numerous, and too numerously attested by independent witnesses, in different ages and countries to be very reasonably denied. It is more rational to believe the facts, consistent as they are to each other, than to conclude in spite of all evidence, that those who relate them are enthusiasts and simpletons.

This latter phenomenon took place even when the eye-lids were kept exactly closed with the fingers.

25. In two somnambulists we found the faculty of fore-seeing the acts of the organism, more or less remote, more or less complicated. One of them announced repeatedly, several months previously, the day, the hour, the minute of the access and of the return of epileptic fits. The other announced the period of the cure. Their previsions were all realized with remarkable exactness. They appeared to us to apply only to acts or injuries of their organism.

26. We found only a single somnambulist who pointed out the symptoms of the diseases of three persons, with whom she was placed in magnetic connection. We had, however, made experiments upon a considerable number.

27. In order to establish with any degree of exactness, the connection between magnetism and therapeutics, it would be necessary to have observed its effects upon a great number of individuals, and to have made experiments every day for a long time, upon the same patients. As this did not take place with us, your committee could only mention what they perceived in too small a number of cases to enable them to pronounce any judgment.

28. Some of the magnetized patients felt no benefit from the treatment. Others experienced a more or less decided relief, viz: one the suspension of habitual pains, another, the return of his strength; a third, the retardation for several months, of his epileptic fits; and a fourth, the complete cure of a serious paralysis of long standing.

29. Considered as a cause of certain psychological phenomena, or as a therapeutic remedy, magnetism ought to be allowed a place within the circle of the medical sciences; and consequently physicians only should practise it, or superintend its use, as is the case in northern countries.

30. Your committee have not been able to verify, because they had no opportunity of doing so, other faculties which the magnetizers had announced as existing in somnambulists. But they have communicated in their report facts of sufficient importance to entitle them to think that the Academy ought to encourage the investigation into the subject of animal magnetism, as a very curious branch of psychology and natural history.

Thus ends the report. An extraordinary impetus was given by it to the study and practice of mesmerism, the name first conferred on these discoveries by Nicolai of Berlin. The names of Colquhon, Mayo, Sandly, Gregory, and many others, are now familiar to all who take an interest in this class of literature. We have not space for a review of opinions, but a short extract from the principal French writer on the effects of magnetism, and Dr. Elliotson's idea of the magnetic influence, will form a suitable appendix to the conclusions contained in the report of the French committee, rendered to the Academy.

When magnetism produces somnambulism, (he says) the being who is in this condition acquires a prodigious extension in the faculties of sensation. Several of his external organs, generally those of sight and hearing are inactive, and all the sensations which depend upon them, take place internally. He has prophetic visions and sensations, which may be erroneous in some circumstances, and which are limited in their extent. He expresses himself with astonishing facility.

We have no right to speak of these but as the result of conditions of common matter. I am satisfied of the truth of clairvoyance, of an occult power of foreknowing changes in the patient's own health that are not cognizable to others—of knowing things distant and things past, and sometimes though rarely, events to come; but I am sure that most clairvoyants imagine much, speak the impressions of their natural state, or of those about them, and may be led to any fancy. Some talk Swedenborgianism, some Roman Catholicism, some Calvinism, some Deism, some Atheism; some prescribe homeopathy, some allopathy; cerebral sympathy, a fact totally unknown to the medical world, is continually mistaken for clairvoyance, and the opinions of the patients may thus be sympathetically those of the mesmerizers. They will deceive from vanity, or love of money or even for fun. Many patients pretend to the power who have it not at all, and those really possessed of it in some cases are not aware of it.

THE PRACTICE OF MESMERISM AND CLAIRVOYANCE.

The recommendation (Report 29) that physicians only should practice, or superintend the use of mesmerism, was sufficiently modest, coming as it did from a body who have had these doctrines forced upon them by public opinion. Of course, the church will continue to recognize the higher facts of clairvoyance, and then it will be very proper that only beneficed clergymen should assume the direction of them!

We do not think, indeed, that magnetism should be practised either *by* all or *before* all; we are too painfully conscious of the abuses to which it is exposed, and of the more subtle injuries that may result from personal contact. These are things that can only be recommended to the serious attention of mothers, husbands, brothers or other friends intimately interested in each case. Certainly, the true magnetism is not a physical, but a mental and religious power, acting however by physical causes. The real power, as we have said, is the *will*, because the will is the momentary gift of the love of God, and it works with his love, when the religious principle is recognized, but otherwise in favor of self-love and self-gratification. The injury done by separating the religious spirit from the magnetic act is incalculable, for it originates in principle all that is properly called "goety" or the "black magic" of the middle ages.

Magnetism is a sacred power, which ought to have but one end, as it can have but one first origin—the elevation of the human race out of their present miseries, spiritual, moral and physical. In the art of magnetizing, the patient is highly sensitive to impressions through the operator, especially to those which flow out spontaneously from the moral ground of his being, and though they may not be felt at the time, they are beyond doubt, *received* as distinctly and more indelibly, than the at first invisible image of the daguerreotype; and a slight cause at some later period, may bring them into manifest existence. We limit ourselves to these hints, and for practical advice concerning the manipulations or process of healing, must refer to the numerous hand-books continually before the public.

The same remarks apply if possible, with greater force to the development of psychological phenomena by magnetism. We have seen performances that were truly diabolical in character and purpose, and the real nature of which may be ascertained by an investigation of principles in the writings of Swedenborg, whom we are here compelled in common honesty to mention. His "Spiritual Diary" commencing when Mesmer was a boy, in 1747 and continued for thirty years, really anticipates the magical process in magnetism by the action of the hand, by breathing, by gestures, words, ideas, and in one place he remarks of certain spirits. These records are, to say the least, curious, considering how they have since been realized on this stage of existence, and without being known, for it was not till after 1840 that the 'Diarium Spirituale' was published by Dr. Tafel of Tubingen, from the original Mss., and by far the greatest portion is still untranslated. The leading principles, indeed, are expressed clearly enough in the works published by the author between 1749 and 1772, but not in their application to those practices which Swedenborg was studious to leave in obscurity.

The French Government did well in these investigations, with the facilities then possessed, but if it could have had what this Government has, or could have today, it would no doubt have made a stronger claim for Vital Magnetism as a remedial agent. The subject is much better understood now than it was in Mesmer's day; and there is opportunity for farther investigation. It is now receiving much attention, and no doubt in the future, will be made productive of much greater practical benefit, when the prejudice against it will necessarily be overcome. This our government should investigate, and place it on an equal foundation with other modes of practice.

VISION OF THE DYING. Are the dying always clairvoyant? Dr. Winslow says "we recollect attending the case of a young lady laboring under a disease which produced extreme mental and physical suffering, who exhibited, a short period before her death, some singular phenomena. This lady had not been seen to smile or to show any indication of freedom from pain for some weeks prior to dissolution. Two hours before she died, the symptoms became suddenly altered in character. Every sign of pain vanished; her limbs from being subject to violent spasmodic contractions, became natural in their appearance; her face, which had been distorted, was calm and tranquil. All her friends supposed that the crisis of the disease had arrived, and that it had taken a favorable turn; and delight and joy were manifested by all who were allowed access to her chamber, and who were made acquainted with the change that had taken place. She conversed most freely, and smiled as if in a happy condition. We must confess that the case puzzled us, and that we were for a short time induced to entertain sanguine hopes of her ultimate recovery. But alas, how fragile were all our hopes! For two hours we sat by the bed, watching the patient's countenance with anxiety. Every unfavorable indication had vanished; her face was illuminated by the sweetest smile that ever played on human countenance. During the conversation we had with her, she gave a slight start, and said in a low tone of great earnestness, "Did you see that?" Her face became suddenly altered, an expression of deep anguish fixed itself upon her features, and her eyes became more than ordinarily brilliant. We replied "What?" She answered "Oh, you must have seen it: how terrible it looked as it glided over the bed!" "Again I see it!" she vociferated with an unearthly scream, "I am ready;" and without a groan her spirit took its flight.

We could fill a large volume with more remarkable cases, which are occurring in our midst. Almost all, immediately previous to their exit from our sphere, to enter upon another life, have glimpses of friends who have preceded them. This experience is so common, indeed, that it is unnecessary to allude to special cases: we cannot doubt, from their frequent occurrence, that they take place in accordance with a law which governs the whole human family.

We read in the Acts of the Apostles (xii. 13.) of the failure and disgrace of certain vagabond Jews, "exorcists," who like the Apostles "took upon them to call over them that had evil spirits the name of the "Lord

Jesus." Their discomfiture was signal, and while it completely disproved their own false pretensions, it as satisfactorily established the reality of the claims of the Apostles to the supernatural power bestowed upon them by their Divine Master.

It is more than probable, however, that this practice among the Jews did not originate from an imitation of the miraculous cures which they had seen performed on the miserable demoniacs, by our Lord and his followers. Traces of another and more ancient source may be observed in a story related by Josephus. "God" says the historian, "enabled Solomon to learn that skill which expels dæmons, which is a science useful and sanative to men. He composed such incantations also, by which distempers are alleviated, and he left behind him the manner of using exorcism, by which they drive away demons, so that they never return. And this method of cure is of great force unto this day, for I have seen a certain man of my own country, whose name was Eleazor, releasing people that were demoniacal, in the presence of Vespasian, and his sons and his captains, and the whole multitude of his soldiers."

Cases like the above are frequently occurring in our midst in the nineteenth century.

With reference to the practical utility of mesmerism, the late Dr. John Elliotson, Professor of Medicine in University College, London, made the following statements in 1840. "Mesmerism is a most useful addition to our remedial means. By it, without giving any medicine, I have several times cured epilepsy; but as the causes of this disease, as of paralysis, are variable, and often irremediable, general success cannot be expected. A case of violent and singular jumping and striking fits of twelve years' duration, lasting weeks, spring and autumn, has yielded to it. A case of chorea was presently cured. A case of intermittent hemiplegia, recurring every few days, and leaving the poor woman in a wretched state for two or three days, has been permanently cured, after resisting all means in Essex for five years. Hysterical insanity, and palsy of the lower extremities, as well as the strangest hysterical symptoms, have yielded to it; and the most distressing hypochondriasis, obstinate hiccup, and that distressing affection termed sick headache, have been cured with it among my patients. Like every other remedy, it can be adapted to a certain number of diseases, and can succeed in a certain number only of modifications of these. The Marquis de Guibert has just published a tract, giving the results of the magnetic treatment, from January 1. 1834 to January 1. 1840. Of 3315 patients, 1194 of whom were men, and 2121 women, the proportion of success is given as follows. Of the men, 663 were cured, 180 relieved, and in 171 the result was not made known; of the women 1285 were cured, 195 relieved, and in 317 the result was not made known."

BIBLICAL ACCOUNT OF VITAL MAGNETISM.

One of the main purposes of this treatise has been to establish the identity of the magnetic processes, and the curative power operating through them, as set forth in the bible, and the same methods and results developed in modern times. In the outset we endeavored to explain the true significance of the cures wrought by the ancient prophets, and those subsequently attributed to Jesus and his apostles and disciples. When viewed in the light of science and philosophy, they are readily comprehended, and entirely compatible with the laws of natural forces, to which they are related. In the sense of wonders simply, the term miraculous as applied to them might be tolerated; but with the commonly understood meaning of something beyond natural law—a setting aside of regular order, and the introduction of a new process which violates the connection between cause and effect, the reasoning mind is confused and unsatisfied. Putting side by side the simple facts as they occurred then and are occurring now, they are found to harmonize throughout, and to be equally explicable by the unerring and invariable operation of the laws of nature.

In the ruder conditions of primeval society, manipulations for healing purposes were performed as religious rites, by persons exercising the duties of the priestly office, and hence when invested with sacred authority, the idea of miracle as an act of special providence, which was associated with them; and this has descended through numerous sectarian channels down to the present day.

In the jewish mode of procedure, the lepers, persons afflicted with hemorrhages, abscesses and ulcerations, as well as puerperal women passing through the period of purification, must be in the personal presence of the priest, who is the principal functionary in the performance of the ceremonies of cleansing. It is reasonable to presume that during these personal visitations and interviews there is a psychological effect produced, exciting an expectation of returning health.

It was stated in a recent lecture delivered in this city that in Constantinople, and throughout Turkey, the Mahommedans are in the constant exercise of the gift of healing. The lecturer witnessed the sudden arrest of hemorrhage in the persons of two howling dervishes, who passed before the Sheik for that purpose. It is the custom of that ecclesiastical functionary, when the sick or wounded present themselves before him, to make passes over them, which it is said, is immediately followed by the healing process.

It is said that in Mahommedan countries disease is believed to have been sent by "Allah" as a punishment for sin; therefore it is accounted presumptuous to attempt to eradicate it; but if a "Hakkom" passes by, the afflicted kiss the hem of his garment with superstitious veneration, and beseech him to lay his hands on and heal them, for every one who has the gift of healing is regarded by them as sent from God, and they often call him "the man of God." They believe he has the power to minister to all their necessities, and that he must be a good man to be endowed with such power. If he refuses to help them (because, it may be, the case is hopeless) they think that "Allah" forbids him; but, on the other hand, if happily he cures them, they thank not him, but "Allah." Thanks, they say, are due only to the Most High; all men are equal in the eye of God, but it is a duty and a privilege to help our neighbors. They show their gratitude to man by acts. They thank God in prayer, and call down blessings on their benefactor; but they have no notion that the art of healing the sick may be acquired or taught.

We find in the annals of history that the Indians for a long time have understood the fact that both good and bad influences were made to operate on the mind through an invisible source. All tribes have their "medicine men," or men of mystery, who in time of disease prescribe various roots and herbs; and if these are not successful, they resort to magic or mystery. These professional men are regarded as worthies of the highest order in all tribes: they are regularly called and paid as physicians. Many of them acquire great skill in the art of medicine, and gain much celebrity in their nation. They also understand the affliction of "possession" either intuitively or otherwise, and when one of their number is troubled by an unruly influence, they seem to understand the conditions requisite to counteract it, and let the captive go free.

Their mode of treatment is to form a ring, and commence playing music for the purpose of producing harmony. Arrayed in a strange and uncouth dress, of fantastic form, constructed according to the wildest fancy imaginable, they make their visits to their patients, dancing and shaking their frightful rattles, and singing songs of incantation, in the hope of relieving the sick person of disease, or counteracting the bad influence. Many strange accounts are given by those who have witnessed their peculiar ceremonies in driving away the spirits from their afflicted ones. Doubtless when they shall have learned the nature of the law under which these things occur, they will improve their mode of practice, and abandon many of their wild gestures.

The Mormons claim the healing power by laying on of hands, among the gifts alleged to be conferred by divine authority, upon certain of their officials.

The bible contains a record of numerous instances in which the power of healing has been exemplified in diseases of both body and mind. The churches as a body receive this record as true, and approve of the mode of cure at that time practised, because of the persons who practised it; but when it is asserted that these things are being done in this our day, and that they can easily learn and know the facts for themselves, they immediately raise the cry of "fanaticism," and declare that they are the works of the "Devil," instead of investigating the matter, and trying to ascertain whether the facts are as represented, and whether the law which governed the cures of ancient times is still in operation.

Magnetizers and others who know the facts should exercise charity towards those in whom a faulty education has engendered a prejudice against the practice. If they are truths, of which we are well assured, time will demonstrate them, and work out the results. Prejudice and unbelief will yield to undeniable proof. Those who are favored with the gift, and those who have seen its usefulness verified, can afford to wait patiently, until evidence can illuminate the minds of the less informed. Many think the gift comes from God by special providence: others assert in a spirit of anathema that it is from his great opponent and adversary, or deny the fact.

It is desirable to be able to assign a reason for every thing, but it is not absolutely necessary that we should know the source from which the power comes. All we need to do is to note the fruits. If they are good, whether they come from the developed, or undeveloped whose goodness is yet in part latent, from the exalted or the lowly, give the credit to the true source. Life's duties require us to perform acts for others as well as ourselves; and what better work can be engaged in than that of relieving the suffering whenever and wherever it lies in our power. Seeing the benefit imparted to the needy ones, we should feel encouraged to persevere in it.

The Romish church has a knowledge of the power of vital magnetism, and as we learn, exercises it at times; and her devotees imagine it to be confined to a few of the leading officials. Doubtless this is the cause of their being so firmly united; but we claim that it is limited to no sect, class, nation or color. It belongs to Jew and Gentile, Catholic and Protestant alike. It is unfortunate that the question of religious belief should ever arise as a barrier in the way of the performance of a good deed to one's fellow man.

The Protestant church ignores the gift of healing in the present age, although full of faith in what is recorded of the works of Jesus and his apostles. But a short time ago we heard a member of a Congregational society say that the church, as a body, had rejected the greatest gift that could be bestowed upon mortals, which she said was that of healing. Said she "I have the gift, and shall use it."

The Methodists in their primitive days were favored with its development among them. The Second Adventists claim this among their spiritual gifts. But the Spiritualists have kept their minds in a state of freedom and rational receptivity, and are more favorably disposed towards its cultivation than any class of religious believers, and are doing more to bring it into use than any of the denominations. This indicates that it does not depend upon mere belief, but that it exists in individuals every where, and has existed either from birth, or from the period of its subsequent development. It is common among the Greeks and Chinese, who have long used it. If it were under the exclusive control of the individuals through whom the manifestations take place, and could be exercised or repressed at will, we should have but little confidence in its being made beneficial to suffering humanity. But as it proceeds from the Author of our being, who is no respector of persons, we have great confidence in its power for good, and hope that ere long there will be few who will look upon it as unnatural, strange or mysterious.

As will presently be seen, by reference to passages familiar to all readers of the hebrew and christian scriptures, Elijah healed the sick in his day, and his mantle fell on Elisha. Others of the prophets were similarly endowed, and so it continued down to the time of Jesus and his apostles, and it continues to this day as was prophesied by Jesus. What better evidence do we need than the identity of the curative results experienced to-day, with those accomplished then, under the operation of the same universal law?

The simple process of laying on of hands, to the superficial thinker seems absurd, and incapable of producing any perceptible or tangible effect. But it has a meaning; and we submit the following, from the pen of a distinguished minister, who many years ago attained the distinction of Doctor of Divinity in an orthodox institution, which he entitles "thoughts on the philosophy of laying on of hands, as a remedial agency in the treatment of disease, suggested by a person in the higher magnetic state."

"They were given from time to time in the course of three years of magnetic treatment under the direction of Dr. J. K. Mitchell, late of Philadelphia. The subject was a son of mine from fifteen to eighteen years of age, who had been an invalid from his birth, from what was judged to be a malformation of the chest, or of some of the organs about the heart. When he was fourteen or fifteen years old, the disease had made such progress as to leave little or no hope that he could live many months longer. Three consultations, one in Boston and two in Philadelphia were held in his case, the result of all which was that there was no course of medical treatment known to the faculty which could be applied to the case with any prospect of relief. In this state of things the magnetic treatment was recommended, not by a physician, but by a lawyer, who had been making experiments in it for his own amusement. The magnetic treatment was applied to my son. It was at once found that he was very highly susceptible to the magnetic influence. A physician who was an expert in the subject of animal magnetism was called in. In a short time he became clairvoyant—could look into the chest, see what the disease was, and from that time the treatment was conducted according to his own directions given in the magnetic state. He would remain in that state sometimes for hours, and in two or three instances for several days—could eat and drink, and journey in it; but what occurred in that state he had not the most distant remembrance of in the natural state. I requested him to write me out an account of the animal magnetism, what it was, and what its influence on the human system. He did so, an outline of which is as follows.

"In the process of magnetizing, there is a fine purple fluid, invisible in the natural state, but distinctly perceptible in the magnetic state, which passes from the magnetizer to the person magnetized. Thus in the case of my son, he said as soon as I began to make passes over him, he could see little threads of purple light come out from my hands and eyes, but more especially from my *hands*, which came to him, and soon pervaded every part of his system. This fluid he said was the *power of life*; that in every healthy person the working of the animal functions produces this fluid or power of life in sufficient quantities for the purposes of life, and in most cases more, so that a healthy person can impart a portion of it to a less healthy one. Thus in the case of my son, he said that in his enfeebled constitution, there was not enough of this power produced—not so much as was expended daily—that when I magnetized him, a portion of my own power of life was passed over to him, which united to what his own system produced, was sufficient to withstand the disease, and finally to overcome it. This power of life is communicated most abundantly through the *hands*, which explains the philosophy of the 'laying on of hands' as a curative agency. And it may be noticed here how the miraculous cures of our Saviour and his disciples were almost always effected through the *hand*. He laid his *hands* on them; he took them by the hand, &c. So when he began his parting charge to his disciples, he said 'these signs shall follow them that believe; they shall have power to cast out devils, and they shall lay their hands on the sick, and they shall be healed.'

"Some persons possess the power of communicating in a much greater measure than others, and some are more susceptible to the influence than others. In the case of my son, the influence was exerted in the magnetic state, but it may be effected, though less freely, in the natural state. According to my theory, in every case where you afford relief by the application of the hand, you impart a portion of your own power of life. There is a law of the human constitution that all our powers become strengthened by exercise; so in this case, as you use this power of life, it will be supplied in still greater measure."

Another minister in a letter addressed to a magnetizer, alludes to the effect of magnetic treatment in his own case, in the following terms: "I send you a few words of cheer, on account of the remarkable cure you have effected in me. May God bless you always, wherever you may be. I write this to you, without any solicitation on your part, in order that you may be encouraged to continue to exercise the mysterious power of healing the sick. Hoping your patients may have as much reason to rejoice in your great success as I have, &c.——"

In the old testament cases, the cures were chiefly confined to the diseases known as leprosy and paralysis, though the character of the disease, as in the case of the widow's son, is not always stated. This case is the first in the historical order. The mother, in her humility and agitated by her maternal affection, connected the child's sickness with some supposable sin of her own, and as a punishment thereof.

The account in the eighteenth chapter of the first book of Kings describes the contact of Elijah's body with that of the child, stretching himself three times upon it, at the same time crying to the Lord to bring back the soul into the body; the breath, pneuma, spirit or soul having apparently departed from it.

Psychological influence was recognized and understood in those days, and the gift, as it may be termed, of prophecy conferred upon particular persons. Its exercise although in general truthful, was in some cases false, precisely as we witness it now.

Elisha asks of Elijah (II. Kings ii. 9 et seq.) a double portion of his spirit, i. e. the gift of healing, and is promised it conditionally on his seeing Elijah when he should be taken away. In the fifteenth verse of the same chapter, its fulfilment is recorded.

In the same Book (iv. 28-37) there is an account of the Shunamite's son who was said to be dead, and Elisha restored him to life. He stretched himself upon the child, who was speechless and insensible, having as the narrative reads, lost both voice and hearing. The circulation was impelled forward, making the body of the child warm again. Elisha returned, walked the house to and fro, his mind of course fixed upon the little patient, in psychological sympathy; and he came again and laid upon him. On the first occasion he put himself in close contact with the child, mouth to mouth, eyes to eyes, and hands to hands, when the latter sneezed and awoke to consciousness.

In the fifth chapter is an account of the cure of leprosy, in the person of Naaman, a military officer of the rank of captain, who had had extensive ulceration. Coming to the prophet Elisha, he was directed to wash seven times in Jordan, when his flesh should come again, and he should be clean. Of course there was no specific virtue in dipping any specified number of times in that particular river, but the psychologic power was exerted through that means. The simplicity of the remedy enraged the aristocratic leper. He expected the name of God to be formally invoked, and as his servants said, some great thing to be done. On being reasoned with, he at length obeyed, "and his flesh came again like unto the flesh of a little child."

But Elisha himself, (ch. viii.) although possessed of such healing power, and a man of God or prophet withal, was not perfect. He prophesied a falsehood in the case of the King of Syria. It is alleged that the Lord had told him that the King should surely die, but the prophet sent word by Hazael "thou mayest surely recover."

In Jeremiah (ch. xxiii.) words of complaint and rebuke are given as the utterance of God himself, against the false prophets "that prophesy lies" from the deceit of their own heart. These and their dreams which they told to their neighbors, and with which they caused the people to forget his name, were the offenses which brought condemnation upon them. In ch. xxviii Hananiah's false prophecy is detailed, for uttering which the Lord declares that he should die during that year. Yet by his psychological power he had made the people to trust in a lie, and at first caused even Jeremiah to say Amen.

In I Kings xxii. 21-23 and II Chron. xviii. 20-22, it is asserted that the Lord put a lying spirit into the mouths of the prophets, that they might entice Ahab, King of Israel.

A cure effected chiefly by prayer, is mentioned in Hezekiah, (ch- xx.) That officer became "sick unto death;" the disease itself is not mentioned, except that he had a boil. The prophet Isaiah told him that he should die. By prayer however, he was cured, a lump of figs being laid on the boil as a physical remedy; and his days lengthened fifteen years. One of those marvellous statements of the interruption of the solar and planetary movements is included in the account—the falling back of the sun's shadow ten degrees, as indicated by the dial of Ahaz, as a sign that he should be healed.

These facts show that the development of spirit power was of very various character, the moral status of the individuals controlled differing widely, as at the present day.

The phase of moral as well as psychological power displayed in the person of Jesus was of a more elevated and practical character than that of the older prophets. It is true that in certain instances he exhibited weaknesses common to humanity, as when he allowed activity to his combativeness, in driving the money changers out of the temple, excited by indignation at their sacrelege; and again in manifesting disappointment at the time of his execution, exclaiming in his agony "My God, my

God, why hast thou forsaken me!" He doubtless believed that, as he said on another occasion, he could command twelve legions of angels to assist him, so God would finally interpose a special power to avert the catastrophe which was the closing scene of the persecutions with which his enemies had followed him. Born with a superior spiritual organization, he was sincere, earnest, self-denying and enthusiastic in the fulfilment of the purposes of his mission. The most essential part of this mission, viz: the exercise of the healing power, continued but about three years, according to the report given us of this eventful part of his life. His birth, the character of his mission, and his death had been foretold through the instrumentality of a power, the efficiency of which is equally operative now, yet denied by many who rely implicitly upon the prophetic testimony of past ages.

He was exceedingly sensitive and sympathetic, taking upon himself the infirmities of others. His movements were watched with intense prejudice, his motives held in constant suspicion, and his aims as far as possible thwarted. He was rebuffed as magnetizers are now, and his cures deferred and hindered in consequence of opposition and unbelief. His spiritual experience was remarkable and severe, and his memorable sayings uttered during the three most eventful years of his career as recorded by the evangelists, are being fulfilled in the present age.

The same enthusiasm, and to a certain extent, the same high moral development is seen in such visionists as Ann Lee, whose psychological power was very great, and remains even now operative among her followers, very much as when she was in earth life.

The doings of Jesus and his disciples are recorded by the several evangelists with varying minuteness of detail, and if the account of a particular event given by one of them be assumed as correct, there will be found more or less inaccuracy in the others. This is necessarily the case where the biographer writes at a period of fifty or sixty years after the events described, and when the information was not always derived from eye-witnesses. When there is a concurrence in the principal particulars, the testimony is regarded as credible.

The circumstance of converting wine into water, which occurred at the wedding feast, is to be explained on psychological principles. It is not uncommon for experimenters in animal magnetism, when their subjects are under full control, to impress their senses in a manner so positive as to equal the effects of the genuine substance.

That the influence of psychological power was known then is obvious from numerous facts. Even as far back as Jacob's time a similar effect was expected from the means which he resorted to for the propagation of animals with particular marks. Strong impressions made upon the mind or affecting nervous impressibility, from whatever cause, belong to this class of influences.

But in this place we have more particularly to do with the power of healing at the beginning of the christian era. It is recorded by Matthew as occurring in the year 27 that Jesus was "led up of the spirit" into the wilderness to be tempted of the devil, (sometimes called evil spirit; we should say undeveloped spirit.) Good spirits came to his assistance, and administered to him. In Hebrews they are called ministering spirits; in the version by Matthew, angels.

He heals "all manner of sickness and all manner of disease among the people." They brought sick people from Syria with "divers diseases and torments," the possessed with devils, lunatics and paralytics. Here a distinction is made between obsession and lunacy; in the parallel passages by the other evangelical writers, as they are termed, the distinction is not observed.

The *possessed* were impressible subjects, inasmuch as the spirits who controlled them held them with a tenacious grasp, until compelled by the superior power operating through Jesus, which he called "the Father," to release them. He did not claim it as his own, but by proper self-discipline, a part of which consisted of fasting and prayer, he kept himself in the best condition of receptivity, highly favorable for receiving in full measure, that wonderful spiritual force, as well as imparting it to others.

The development of the power in the persons of the apostles, and its exercise through them took place in the same way. They had the same fiery furnace to pass through, which characterizes the experience of the magnetizers of to-day. Paul's development was attended with much difficulty. The nature of the necessary course of preparation was not comprehended by him, nor by his co-laborers. Many did not know that it had lain latent in them until quickened into a living force by the vitalizing energy of Jesus.

Paul had strenuously opposed those who exercised the gifts, but being arrested by a startling vision, falling into a state of trance, his mind was enlightened as to their reality, and he was irresistibly impelled to change his course from persecution to zealous advocacy. This change has been designated conversion to christianity. Conscious of spiritual impressions and experiences, when speaking of visions and revelations, in his second letter to the Corinthians, he says he knew a man, but whether in the body or out, he knew not, and that such an one was caught up into the third heavens, and heard unspeakable words, and so on.

The undeveloped spirits controlling the cases above referred to, had a knowledge of the power vested in Jesus, and spoke with an audible voice, expressing their unwillingness to leave their victims or subjects. The plural is used in regard to the influence controlling one particular person. "We are legion"—a combination of spirits. "My name is legion." One individual spokesman answers for all. In Mark the personage here particularized as "devil" is called an unclean spirit. The parallel passages may be found in Mark ch. i. and iii. The idea so commonly entertained in christendom of the existence of a particular being, having a character of unmitigated evil, with power not only coextensive with that of an omnipotent God, but capable of transcending it, and of thwarting his designs and purposes in relation to the ultimate destiny of a vast majority of mankind, has been the means of causing a total misapprehension of the subject in many sincere and truth loving minds. The mere name "devil" is synonymous with *demon*, and demon understood in its legitimate signification is by no means an opprobrious name. The names messenger and angel in those days meant the same thing. Demon or familiar spirit was originally oftener used to indicate a kind and friendly being, than one indescribably malignant. But demons were both good and bad, or more properly speaking, developed and undeveloped. The degeneration of the term to what is now understood as diabolical, or infinitely depraved is the offspring of perverted ideas.

In the eighth chapter of Matthew it is related that a man afflicted with leprosy came and worshipped Jesus, declaring his faith in the healing power. On being touched he was cleansed, and directed to show himself to the priest, according to the formality of the Israelitish law.

Then came a centurian at Capernaum, beseeching him to heal his servant who had paralysis and was "grievously tormented." The mode of expression is singular as applied to a case of loss of sensation and muscular power. The cure was performed at a distance, as a reward, seemingly, for the confiding faith of the Roman officer. Such great faith had not been found, "no, not in Israel."

Peter's mother-in-law was cured of a fever by her hand being touched. In the evening many cases of obsession were brought, and the spirits cast out "with his word." All the sick were healed on that occasion.

In the country of the Gergesenes he was met by two obsessed persons, whose symptoms had been those of violent mania. They came out from among the tombs, and the spirits spoke, beseeching that if they were driven out, they might be permitted to enter a herd of swine feeding at a distance, which was granted, resulting in the drowning of the animals.

In the ninth chapter, another case of paralysis is mentioned, the patient being brought on a bed. His sins were pronounced forgiven, though what particular sins caused the paralysis is not stated. Doubtless he had transgressed the laws of hygiene, as most persons do. The by-standers charging his benefactor with blasphemy for uttering that expression, the paralytic was commanded to arise, take up his bed, and go to his house, which, having the power given him to do, he obeyed.

A ruler asks him to visit his daughter, although he thinks she is already dead. Yet the ruler's faith is strong, believing that if the great healer will lay his hand upon her, she will live. On the way, a woman who had suffered from hemorrhage of twelve years' duration came behind and touched the hem of his garment, in the firm belief that thereby she would be cured. In Mark's testimony the statement contains additional particulars. Jesus was sensitive to the magnetic movement—"virtue had gone out of him." The poor nervous woman feared and trembled, lest she should have done something wrong, in her imaginary presumption. The cure was effected, and stress was laid upon her faith in securing it.

On reaching the ruler's house, he pronounced the daughter not dead, but asleep. They treated his assertion with contempt and ridicule. He took her by the hand, however, and she arose.

Two blind men came. He touched their eyes, and restored their sight. Here again faith is exercised, as if it was recognized as a fundamental condition. A man who was dumb and obsessed was brought to him. In this case, the mute condition depended upon the obsession, for on the expulsion of the disturbing spirit, speech was restored. Then it was that the pharisees charged Jesus with driving off these undeveloped spirits through the prince of them, Beelzebub or Baal-zebub, the interpretation of which is unimportant.

In the year 30 of the christian era, the twelve disciples were called together, and the healing power imparted to them, with instruction to dispense it freely. "Freely ye have received, freely give." (ch. x.)

In ch. xi. the result is given; "The blind receive their sight, and the lame walk, the lepers are cleansed, and the deaf hear, the dead are raised up, and the poor have the gospel preached to them." This proclamation was indeed gospel or as the word inplies, "good news" to the sick and suffering poor.

In ch. xii. there is a case of paralysis or atrophy of the hand, which on being stretched forth in obedience to the direction of Jesus, was restored whole like the other. It was on this occasion that the malignant pharisees sought to entrap him with their sabbatarian prejudice; but he silenced them by the use of an argument in defence, based upon their own practices. He inquiringly appealed to themselves, to know if any one of them having a sheep which had fallen into a pit on the sabbath day, would hesitate to lift it out; and asked how much better is a man than a sheep? He showed that it was not only lawful, according to the letter and spirit of the Israelitish code, to do good on the sabbath day, but right and proper in itself.

We have here also another case of obsession, wherein both speech and sight were lost; and again on the patient being cured, the same pharisees repeated the sneering assertion as to the source of his power, and he again rebuked them into silence by an unanswerable argument. He cited them to their own conduct, illustrating by those well-known similitudes, "The tree is known by his fruit," and "Out of the abundance of the heart the mouth speaketh."

The obstinacy of the people in his own country, in refusing to hear and understand, follow in ch. xiii. In the latter part of this chapter, it is stated that the people were astonished at his wisdom and his mighty works. Here also is the celebrated saying "A prophet is not without honor, save in his own country, and in his own house." One writer adds also "his own kin." Luke has it, "is not accepted." He worked less there because of their unbelief.

His compassionate nature toward the multitude is exercised, (ch. xiv.) and as many as came in from the country round about were, by touching the hem of his garment, made perfectly whole.

In the succeeding chapter, a Canaanitish woman beseeches him to relieve her daughter, who is grievously vexed—an obsession case. He at first makes no answer. The disciples were unsympathizing, because of her nationality, and desired that she should be sent away. He replies that he was sent only to the lost sheep of the house of Israel, and alludes to the selfish prejudice which accounted the people of her nation as dogs; but in her humility she pleads again, and in consequence of her great faith her prayer is granted, her daughter being made whole from that very hour. It is a question how the faith of the mother could be made to affect psychologically, the condition of the daughter.

Great multitudes of persons came, bringing with them those that were lame, dumb, blind, maimed, and many others, all of whom were healed.

Chapter xvii. contains an account of an insane young man, "sore vexed," as it is quaintly stated. The case was probably complicated with epilepsy, as the patient frequently fell into the fire, and into the water. Jesus rebuked the "devil," and he departed out of him. The condition of lunacy, as in the beginning of the statement it is termed, is here used synonymously with obsession. The disciples had failed to cure him, and on being upbraided for their failure, they asked privately, (apart) why they had not succeeded, and were answered, "because of your unbelief," and were assured that if they had faith, as a grain of mustard seed, nothing should be impossible to them. The diseased condition that distinguished this case was not, however, to be removed but by prayer and fasting, showing that there was a proper physical and spiritual state of preparation which was essential to success. Jesus had paid attention to diet and other hygienic requirements, which in so far placed him in that condition, but these the disciples neglected.

In ch. xix. it is stated that great multitudes followed

him, and he healed them there. In ch. xx. two blind men sat by the way-side, and importuned for mercy. The multitude rebuked them, but Jesus had compassion on them, and touched their eyes, when they immediately received their sight, and followed him.

The last of the series of cures ralated by Matthew, were performed on the occasion of his entering the temple, (ch. xxi.) when the blind and the lame came in, and he healed them. He taught at the same time, and then it was that the chief priests and elders of the people came to him and asked him by what authority he did these things. But he was clairvoyant and psychometric, by which means he was enabled to read their thoughts, as he had previously done those of the crafty pharisees. He asked them a question in return, concerning the baptism of John, which they were afraid to answer, when he also declined to tell them the authority by which he performed his cures.

Mark in ch. x. gives circumstantially the case of the blind man Bar-timeus, whose faith was strong, so much so that Jesus on restoring his sight, declared that it was that which had made him whole.

In ch. xvi. it is stated among the signs which should follow those who believed, that they should lay their hands on the sick, and they should recover; and the relief of obsessed persons is mentioned in the language peculiar to the New Testament writers.

Luke (ch. iv.) relates that Jesus, on coming into Nazareth, entered the synagogue, and as was his custom, stood up to read, when the book of the prophet Esaias was handed to him, from which he read a passage beginning with the words "The spirit of the Lord is upon me, because he hath anointed me—" Among the things to be accomplished by this mission, for which the figurative expression "anointed" is used, was the restoration of sight to the blind, and healing the broken hearted. On closing the book and returning it to the minister, he said "This day is this scripture fulfilled in your ears."

Chapter six contains the account of healing on the sabbath, already quoted, and also the selection of twelve persons from among the disciples, whom he named Apostles, and invested with healing power.

In the eighth chapter, it is recorded that among certain women who had been healed of evil spirits and infirmities, Mary, called Magdalene, had seven of such spirits, which went out of her.

John (ch. xi.) gives an account of the raising of Lazarus to life. The body had been placed in a cave, but not fully interred. In the early part of the account Jesus says "This sickness is not unto death." He had already said in the eleventh verse "Our friend Lazarus sleepeth; but I go that I may wake him out of sleep." Afterwards he says plainly that Lazarus is dead. The terms "dead" and "half dead" are variously employed in other portions of the scriptures. It is a remarkable fact that a case of so much importance should not have been noticed by the other writers. We have elsewhere discussed the question of apparent death.

In ch. xiv., v. 12, Jesus says that he can be excelled in healing. "The works that I do, shall he do also; and greater works than these shall he do."

In the book entitled Acts of the Apostles, and onward, we have examples of the power of healing exercised by some of the apostles, fully as astounding as the acts of their teacher. Peter (ch. iii.) cures a lame man, forty years of age, whose deformity was congenital. He is assisted by the spirit of Jesus, in whose name the cure was effected. After this time, it is related, sick people were brought forth into the streets, and laid on beds and couches, that at least the shadow of Peter passing by, might overshadow some of them. Others were brought from the cities around Jerusalem, some of them "vexed with unclean spirits," and every one healed.

In the eighth chapter, similar cases are mentioned. The excitement caused by these deeds, aroused the cupidity of a certain man named Simon, a sorcerer, who had performed numerous acts of magic, and evidently wielded great psychological power for a long time, insomuch that the people said "This man is the great power of God." He offered to buy the gifts with money, but Peter rebuked him with severity, for his unholy desire.

Ananias, (ch. ix.) in obedience to instructions received in a clairvoyant vision, enters a house where Saul of Tarsus was tarrying, and who was then afflicted with blindness. Putting his hands on Saul, and delivering his message, sight was restored, and immediately there fell from the patient's eyes something resembling scales, which, of course rendered his sight clear.

In the same chapter we are informed that Peter there found Eneas, a paralytic, who had kept his bed eight years, who on being commanded to rise from, and make his bed, rose immediately.

Next comes the bringing to life of Tabitha or Doreas at Joppa; a woman beloved for her good works.

An account of the cure of a case of congenital lameness is given in the fourteenth chapter, in which the efficiency of faith is again prominent. The man had never walked, but now both walked and leaped.

Handkerchiefs and aprons were magnetized by being placed in contact with Paul's body, which when brought to the sick, cured their diseases, and expelled the turbulent spirits from the obsessed. The sons of Sceva, and other vagabond Jews attempted to exorcise spirits, calling the name of Jesus, but having no power, were punished for their temerity, by the obsessed persons leaping upon and injuring them. (Ch. xix.)

An accident befel a young man called Eutychus, who sat in a window, and falling asleep during Paul's long preaching, lost his balance, and was precipitated from the third loft, of course inflicting serious injuries upon his body, so that it is said that he was taken up dead. Paul, however, ascertained that life was still in him, and his friends were not a little comforted; the inference being that he was restored by the apostle, although it is not so stated, the account being incomplete.

The father of Publius being sick of a fever, and having, superadded to this, suffered from an attack of dysentery, in the island of Melita, where Paul had been shipwrecked, the apostle prayed, and laid his hands on the sick man, and healed him. When this was done, others in the island, came and were cured of disease.

These events took place at a period extending from the time when the active career of Jesus closed, continuously to the year 61. On carefully examining the record, the reader cannot fail to be struck with the similarity of events constantly transpiring about us to-day.

The diversities of spiritual gifts are enumerated in I. Corinth. ch. xii. et seq., and in each instance, the gift of healing stands equally prominent with the others; and in the fifth chapter of the Epistle of James, it is enjoined upon the sick to call for the elders of the church, who, together with their religious ceremonies, are required to make use of physical means, and have had set before them numerous examples of the successful application of vital magnetic treatment for healing purposes.

CONCLUSION.

In concluding this brief treatise, it is only necessary to remark that it is an attempt to discharge a duty, to which we have been impelled for a number of years. An important subject required to be more definitely set forth than in detached fragments, as heretofore. Relying upon a plain statement of historical facts, we have essayed to show that the vital magnetic power, known under the various names presented at the outset, has been exercised almost from time immemorial. Recognizing no authority as positive and absolute, we do not ask the reader to accept any theory which does not commend itself to his common sense. We have been prompted by a desire to benefit humanity; a vast field, in which it is the duty and privilege of all to labor.

"Truth is eternal, like divinity itself. Man may neglect it for a long time, but the moment comes at last, when in the fulfilment of the decrees of Providence, its rays pierce the cloud of prejudice, and throw over mankind a beneficent light, which henceforth nothing can extinguish."

"God holds the same relation to the Universe, and its entire outgrowth, as the spirits of human beings do to the body material."

In quoting from the Bible records, the history of cures, and bringing in comparison with them similar occurrences taking place much more frequently in modern times, we have not been actuated by a spirit of unfair criticism, nor do we wish to detract from the sacredness of other portions of the book; but our object has been rather to show that the power which produced the one series, is identical with that which has produced the other. Throughout human history, all proceed from the same source—truth lies at the foundation; and while all that has been presented is truthful, all is natural.

Finite minds have not comprehended, nor can they now comprehend, to the fullest extent, the *universal life force* that has always existed, continues in the present day, and will through generations yet unborn, in the future ages. Phenomena which have been considered miraculous, seemed so, only because of ignorance of the law under which they were developed.

Reposing confidence in the candor, and progressive spirit of our readers, we commit this little volume to their unbiased consideration, in the hope that it will be the means of restoring many to health, who would otherwise remain in the vale of affliction, and thus contribute its mite to the great work of humanity.

www.ingramcontent.com/pod-product-compliance
Lightning Source LLC
Chambersburg PA
CBHW021705210326
41599CB00013B/1531